RAND NATIONAL DEFENSE RESEARCH INSTITUTE

T0122927

Capabilities for Joint Analysis in the Department of Defense

Rethinking Support for Strategic Analysis

Paul K. Davis

Prepared for the Office of the Secretary of Defense

For more information on this publication, visit www.rand.org/t/RR1469

Library of Congress Cataloging-in-Publication Data
ISBN: 978-0-8330-9548-0

Published by the RAND Corporation, Santa Monica, Calif.

© Copyright 2016 RAND Corporation

RAND® is a registered trademark.

Cover image adapted from iStock/hh5800.

Support RAND
Make a tax-deductible charitable contribution at
www.rand.org/giving/contribute

www.rand.org

Preface

This report is a response to Section 1053 of the National Defense Authorization Act for Fiscal Year 2015 regarding the Department of Defense (DoD) budget; the section called for an independent study regarding DoD's joint analytic capability and recommendations for improving it. This report is intended primarily for policymakers and staff in DoD and Congress. The report may also be helpful to the much broader community interested in analysis for strategic planning, whether in government or the private sector, and related methods and tools. Comments and questions are welcome and can be addressed to the author at pdavis@rand.org.

The research for this report was conducted for the Office of the Secretary of Defense within the International Security and Defense Policy Center of the RAND National Defense Research Institute, a federally funded research and development center sponsored by the Office of the Secretary of Defense, the Joint Staff, the Unified Combatant Commands, the Navy, the Marine Corps, the defense agencies, and the defense Intelligence Community.

For more information on the International Security and Defense Policy Center, see www.rand.org/nsrd/ndri/centers/isdp or contact the director (contact information is provided on web page).

Contents

v

Figures and Tables

Figures

Tables

Summary

Background

The U.S. Department of Defense (DoD) has long had broad capabilities for joint analysis, whether for planning future forces or supporting field commanders during current challenges. DoD is envied by other government agencies for its analytic capabilities and processes. Nonetheless, some issues have arisen. This report stems from a congressional request for an independent report about DoD's capabilities for joint analysis and ways to improve them. Congressional concerns largely involved the activity called support for strategic analysis (SSA). In 2011, the director of the Office of the Secretary of Defense's (OSD's) Cost Assessment and Program Evaluation (CAPE) disbanded campaign-modeling and reduced CAPE's participation in the SSA activity, weakening SSA significantly. This report is largely about SSA and how and whether to revise it.

Differing Views

CAPE's decision reflected a conclusion, accepted by the Secretary of Defense and some other senior leaders, that the SSA process had not helped decisionmakers confront their most-difficult problems. The activity had previously been criticized for having been mired in traditional analysis of kinetic wars rather than counterterrorism, intervention, and other "soft" problems. The actual criticism was broader: Critics found SSA's traditional analysis to be slow, manpower-intensive, opaque, difficult to explain because of its dependence on complex models, inflexible, and weak in dealing with uncertainty. They also concluded that SSA's campaign-analysis focus was distracting from more-pressing issues requiring mission-level analysis (e.g., how to defeat or avoid integrated air defenses, how to defend aircraft carriers, and how to secure nuclear weapons in a chaotic situation). CAPE felt that the focus on analytic baselines was reducing its ability to provide independent analysis to the secretary. The campaign-modeling activity was disbanded, and CAPE stopped developing the corresponding detailed analytic baselines that illustrated, in detail, how forces could be employed to execute a defense-planning scenario that represented strategy. During the secretary's reviews for fiscal years 2012 and 2014, CAPE instead used extrapolated versions of combatant commander plans as a starting point for evaluating strategy and programs.

The CAPE decision was controversial. The services all formally expressed their concerns because they value the SSA process and believe that missing aspects of SSA should be restored so that joint analytic baselines will be available to use for the services' analyses (e.g., analysis of alternatives [AoAs] and program development over time). The decision was also controversial within OSD. Some senior officials believed from personal experience that SSA had been very useful for behind-the-scenes infrastructure (e.g., a source of expertise and analytic capability) and essential for supporting DoD's strategic planning (i.e., in assessing the executability of force-sizing strategy). These officials saw the loss of joint campaign-analysis capability as hindering the ability and willingness of the services to work jointly. The officials also disagreed with using combatant commander plans instead of scenarios as starting points for review of midterm programs, because such plans are too strongly tied to present-day thinking.

The reader will appreciate, then, that disagreements are strong across and even within DoD offices. The conclusions and recommendations on the future of SSA are based on interviewing, selective review of government documents, and prior research. Although idealized, the proposed approach is feasible. It builds on the many successes of the past SSA and on modern methods of analysis.

Conclusions and Recommendations

Research for this report substantiated *both* the criticisms that led to CAPE's actions *and* the need to recreate capabilities that have been lost, albeit in different form. What should be done, however, is not straightforward, and halfway measures could be counterproductive.

Background for Recommendations: Reconceiving the Larger Planning Construct

Some of the root problems affecting SSA stem from how DoD conceived SSA years ago. Figure S.1 is a simplified version of the approach as it existed through 2011 for midterm planning.

At first glance, this pre-2012 approach seems reasonable. In this construct, the Office of the Under Secretary of Defense for Policy (OUSDP) issues the defense-planning scenarios based on defense strategy, the Joint Staff specifies concepts of operations (CONOPS), and CAPE then creates a study to specify enough contextual detail to justify assumptions and other data necessary to feed campaign models. CAPE reports the package as an analytic baseline consisting of a scenario (a written description), CONOPS, integrated joint data, and model results. This baseline allows competing analyses to be compared with common data and assumptions, thereby making evaluation easier. The baseline also provides initial data for subsequent analyses, such as AoAs and joint studies.

Upon reflection, the larger construct of Figure S.1 has serious problems. First, it omits the most important step: determining the defense strategy and budget in the first place. Does that not need resource-informed analysis (in addition to background research)? Second, the figure shows planning scenarios flowing out of strategy. Does that not need analysis? Third, in practice, the CONOPS developed to execute the scenarios end up specifying a much more precise scenario than intended by the descriptive scenarios generated by OUSDP. Whereas OUSDP's scenarios depict numerous difficulties, dilemmas, and uncertainties—sketching a rich problem *space*—the CONOPS are developed to describe a narrow path through the space.

Figure S.1
Simplified Depiction of the Support for Strategic Analysis Process (Through 2011) for Midterm Planning

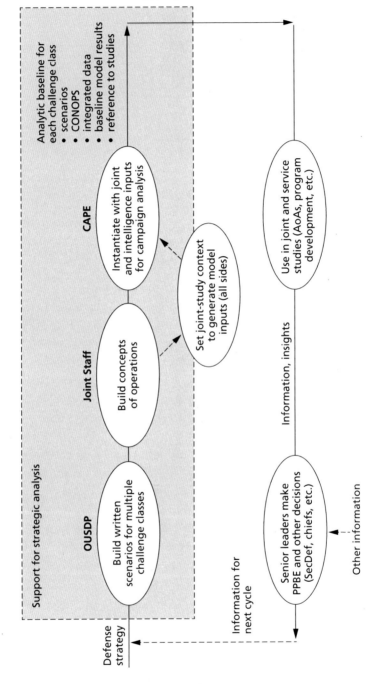

NOTE: PPBE = planning, programming, budgeting, and executing; SecDef = Secretary of Defense; AoA = analysis of alternatives.

RAND *RR1469-S.1*

Although some excursions can be performed, many crucial uncertainties and disagreements are lost.

This lack of excursions was not intended: Analytic baselines were to be mere points of departure for far-reaching excursions, but except for some included with the baseline packages, such excursions proved rare because—given SSA methods—they required new CONOPS and other new data, the development and coordination of which were a major undertaking. The resulting analytic process buried uncertainty and disagreement. The process was unable to adequately assist senior leaders when they sought to address uncertainty, what-ifs, and trade-offs. The process was also lengthy and tedious despite admirable cooperation, expertise, and professionalism.

Ironically, the Secretary of Defense introduced capabilities-based planning in 2001, after a decade of recommendations about dealing better with uncertainty. Successive secretaries have reinforced that intent, which translates into a desire to seek capabilities providing flexibility, adaptiveness, and robustness (FARness) to the extent permitted by a budget.[1] The analytic process, however, has seriously undercut such intentions. *A root problem has been that the level of strategic analysis has been too detailed, precluding more far-reaching analysis.* That detail has also generated objections from the Joint Staff and services, who feel that the baselines have overly specified crucial matters that should be addressed contextually by the military.

Another objection to the SSA activity, both before and after 2011, is that it has accepted uncritically the services' programs of record and traditional CONOPS. This is a particular problem when defense secretaries are deeply concerned about such adverse trends as increasingly capable adversaries with better operational concepts and seek to take advantage of new technologies with innovative concepts of organization and operations, rather than rely on extrapolations from the past.

A final background observation: The pre-2012 SSA activity was conceived of as providing standardized models and data (including assumptions) for subsequent detailed campaign analysis. Common joint data on factual matters are important, and having a reference set of common assumptions is useful when comparing analyses, but the near-exclusive de facto focus on standard models and standard inputs has been counterproductive, defeating the intentions of planning under uncertainty. The pre-2012 activity created substantial overhead expense, with questionable analytic value to leadership. Revitalized analytic capability is needed, but—if the activity is to be more useful—DoD should establish a base of uncertainty-sensitive analysis on which decision-oriented analysis can draw when necessary—e.g., well-conceived trade-off charts relating both to objectives and to risks. The recommendations described below would accomplish that, among other things.

A New Planning Construct: Analytic Support for Strategic Planning

Against this background, this report recommends that DoD should create a new activity to replace SSA. The new activity could be called *analytic support for strategic planning* (ASSP).

1. ASSP should support *initial, interim* decisions on defense strategy and budgets with explicit, understandable, and therefore relatively simple analysis.
2. ASSP should emphasize planning for flexibility, adaptiveness, and robustness (FARness); it should deemphasize detailed analytic baselines. As occurs now, it should develop a list of type scenarios (e.g., defeat a particular rogue state that is attacking an ally, prevail in two simultaneous wars, and cope adequately with even more-complex cases).[2] Then, *for each*, ASSP should develop spanning sets of variations to stress U.S. capabilities in all the dimensions needed. The spanning-set scenarios should constitute necessary and sufficient requirements for the services to meet in their program development.
3. ASSP should include options that incorporate emerging technology and innovative concepts. To assist in doing so, DoD should elevate the role of the Office of the Under Secretary of Defense for Acquisition, Technology and Logistics (USD[AT&L]) and have the Joint Staff be more active in assuring the surfacing of innovative service options (not just programs of record and traditional CONOPS).

4. ASSP should help OSD focus on strategic considerations and relatively low-resolution analysis when establishing planning scenarios or studying mission-level issues (also called *capability issues*), with the Joint Staff having the primary role for the next level of detail that is especially important to the services.
5. DoD should require the Joint Staff to be more active in reviewing and critiquing service programs and requirement estimates.
6. ASSP should change the mix of analytic methods, increasing the emphasis on lower-resolution analysis with relatively simple qualitative models, quantitative models, *and* human wargaming. These should be sound abstractions from more-detailed (higher-resolution) work. ASSP should apply this mixed-method approach to both campaign- and mission-level analyses.

Figure S.2 schematically depicts this construct for mid- and longer-term planning. The lightly and more darkly shaded portions indicate what would fall within ASSP (with the option of considering the activity of the colored oval as separate, since it is accomplished by OSD and the Joint Staff alone). As discussed more fully in the text, *implementation would require significant and difficult changes of organization, process, culture, methods, and skill mix.*

Early Decisions with Scenario Sets as Outputs

The first step in Figure S.2 is to inform initial decisions about strategy and budget. The primary issue is what challenges the United States will take on in developing its defense program—for example, does the country want the ability to fight and win two near-simultaneous wars with regional adversaries, that plus deterring several others, or something more complex? The more challenges the United States takes on, and the higher the goals set for each, the higher the budget that is necessary. Some goals are implausible even with very high budgets. Informing such decisions requires analysis that is understandable, broad in scope, and helpful in making strategic choices despite deep uncertainty. Choices need to strike a balance across objectives and risks.

If the decisions are to determine the requirements for the military services, then what requirements are needed, and how should

Figure S.2
A Process for Analytic Support for Strategic Planning (ASSP)

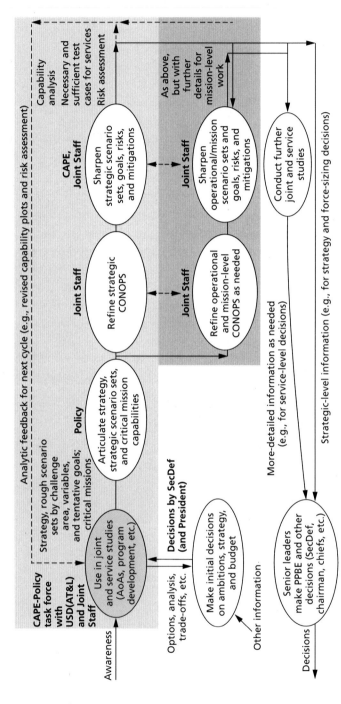

NOTES: Light and dark gray areas indicate the domain of ASSP; work in the darker area is somewhat more detailed. Policy = Office of the Under Secretary of Defense for Policy; PPBE = planning, programming, budgeting, and executing; SecDef = Secretary of Defense.

RAND RR1469-S.2

they be expressed? Using concrete scenarios is useful and traditional for DoD. The issue is deciding which and what kind. A major change in the 1990s was DoD beginning to use a number of type scenarios (e.g., defeating a regional rogue, defeating a near-peer regional power, and simultaneous conflicts). DoD has since added more-complex combination challenges (e.g., dealing with attacks on the homeland while being engaged abroad in two theaters). Each type scenario could arise in diverse ways, demanding different capabilities of U.S. forces. A principle for evaluating a force-planning strategy is that analysis must cover the variations: It is having the capability to do well *across the range* of reasonable possibilities that matters, not just having the capability for some base cases.

As demonstrated in modern work, dealing with myriad variations coherently can be accomplished using a well-chosen spanning set of scenarios; capability to deal with those will imply capability to deal with in-between or lesser-included possibilities. This is not planning for the so-called worst case, because no such worst case exists. A case that is the worst in some respects (e.g., short warning) will be easy in others (e.g., the adversary will also have fewer forces than after a mobilization). Consequently, if requirements are to be expressed as test-case scenarios, many such cases (a spanning set) are needed for each type scenario. Defining these is straightforward for analysts, but not so for committees. Systematizing such work will be a natural but important extension of the analysis for the 2010 Quadrennial Defense Review, which developed different spanning sets to test the overall force structure by stressing the forces in different ways.

Contrasting Old and New Approaches

Figure S.3 contrasts old and new approaches. Under SSA, an analytic baseline was a single rather detailed example (the ovals) of a type scenario or challenge. In the new approach, each such challenge type is represented by a *set* of low-resolution cases (the circles) presenting different stresses for U.S. capabilities.[3] Because of the simple low-resolution character of the cases, it is possible with suitable models to examine analytically all of the cases and to include parametric variations. As illustrated schematically on the right side of Figure S.3, results of such

Figure S.3
Single Type Scenarios, Spanning Sets, and Capability Maps

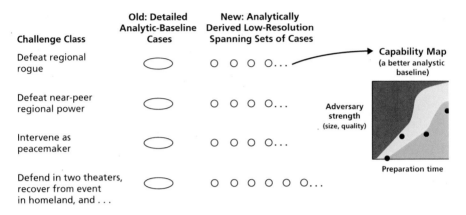

NOTES: Green, yellow, and red correspond, respectively, to probable success, uncertain results, or probable failure. Bullets indicate test cases, which constitute the spanning set.
RAND *RR1469-S.3*

exploratory analysis can be presented in *capability maps* showing for which cases capabilities are *sufficient*, adequate for deterrence at best, or inadequate. That is, green, yellow, and red correspond, respectively, to probable success, uncertain results, or probable failure. Good quantitative analysts can develop these or analogous maps based on broad considerations and low-resolution models; the analysts can then identify the test cases (indicated by bullets) that stress the future force in the different ways needed and that constitute the spanning set. Those sets can be studied in more detail, improving and enriching the capability maps. These selective, detailed looks are sometimes crucial.

Why is this important? Why not just have one illustrative scenario of each class and assume that the services will build capabilities for all the variants? Arguably, during the Cold War, U.S. defense budgets were large, prices were lower, and the services included just-in-case capabilities and slack. That is no longer the case given the budgetary strains on the U.S. military as noted with alarm in the 2014 Quadrennial Defense Review. DoD needs to be more explicit about the breadth of requirements, or the capabilities might not be programmed.

Although the colored capability map in Figure S.3 is simplified and has only two dimensions of uncertainty (preparation time and adversary strength), the basis for more-complex versions of such analysis has been laid over the past two decades with concrete examples. The approach is not hypothetical and can be comprehensibly extended to more low-resolution dimensions of uncertainty (e.g., aggregate-level characterizations of the behavior of U.S. allies and adversary allies, the operational strategies adopted, the effectiveness of new weapon systems, and the effect of such shocks as initial cyber attacks on command and control). Although now well demonstrated, such analysis requires different methods and models than those used in the SSA activity, as well as people skilled in using the new methods and models.

Balancing Mission- and Campaign-Level Analyses

Another objective for an ASSP is elevating the relative emphasis on mission-level work (also called *capability-area work*). Many crucial defense-planning issues are about how to achieve capabilities for accomplishing such increasingly difficult missions as defeating integrated air defenses. The ASSP activity should achieve a balance that gives at least as much, if not more, attention to mission-level issues. OUSDP and CAPE should be concerned primarily with higher-level (lower-resolution) aspects of both, while the services must address such matters in more depth. Precisely the same methods as described above apply to mission-level analysis.

Despite the challenges in doing so, ASSP should also achieve a balance with respect to types of conflict. In particular, ASSP should give appropriate weight to analysis relating to counterterrorism, stabilization, and irregular warfare. The methods for studying these matters should draw on social science (and, in some cases, be the methods of social science).

Responsibilities

As summarized in Table S.1, the new construct indicated in Figure S.2 changes the responsibilities of DoD components in important ways.

Table S.1
Responsibilities in the Analysis for Support of Strategic Planning Activity

	Inform Initial Decisions on Strategy and Budget (Task Force Co-Led by Cape and OUSDP)	Articulate Strategy, Challenges, and Spanning Sets for Challenge	Refine Strategic CONOPS	Refine Strategic Capability Requirements and Assessments of Capability and Risk	Refine Detailed Requirements and Assessments of Capability and Risk as Needed
OUSDP	●●●●	●●●●	●	●	●
Joint Staff	●●	●	●●●●	●●	●●●●
CAPE	●●●●	●	●●	●●●●	●●
USD(AT&L)	●●	●	●●	●●	●●
SecDef	Approval	Approval		Approval	
VCJS, J5, or J8			Approval		Approval

NOTES: The task force activity (left side) might or might not be considered part of ASSP per se. Number of bullets indicates relative responsibility. SecDef = Secretary of Defense; VCJS = Vice Chairman, Joint Staff; J5 = Director for Strategic Plans and Policy, Joint Staff; J8 = Director, Force Structure, Resources, and Assessment, Joint Staff.

The first step of developing defense-planning scenarios is replaced by the larger step of informing interim decisions on strategy and budget. This should be accomplished by an elite team or task force (not a committee) led jointly by CAPE and OUSDP—CAPE because the effort requires rigorous (albeit low-resolution) analysis involving capabilities, costs, and trade-offs and OUSDP because the effort is about establishing strategy. DoD should consider having some top-quality analysts *functionally* dual-hatted for CAPE and OUSDP (even though the analysts would formally reside in only one). DoD used an approach of this sort in earlier decades. Representatives from USD(AT&L) and the Joint Staff should ensure attention to innovation and military feasibility. Results should be approved by the Secretary of Defense.

The next step shown in the table is that OUSDP and the Joint Staff should, as before, articulate strategy and develop CONOPS, respectively. The last two columns are new, corresponding to the lightly and more darkly shaded regions on the right side of Figure S.2. In this construct, the more strategic, low-resolution activities are primarily the responsibility of CAPE, with the Secretary of Defense approving the final expression of requirements. Those of a more detailed nature are primarily the responsibility of the Joint Staff. Although all ASSP contributors would review and comment on everything, as occurs within SSA, Table S.1 indicates that the Joint Staff and USD(AT&L) should be involved in the low-resolution refining and that CAPE and USD(AT&L) should be involved in the more detailed work. The goal should not be consensus but ensuring quality and innovation.

The later portions of the ASSP process (the part most akin to the current SSA) would have new responsibilities for refining capability maps and related risk assessments. Although ASSP would do supporting analysis (e.g., the creation of capability maps), the analysis directly aiding decisionmaking would continue to depend on CAPE (for the Secretary of Defense) and the various special groups serving the chairman and service chiefs. ASSP would still be providing support, not final analysis, but its products would be far more informative than data for point cases.

In this construct, CAPE would reassume some of the role it had in SSA, but with crucial differences. CAPE would focus primarily on the strategic level rather than be dragged into more detail, and it would be refining higher capability requirements at campaign and mission levels rather than specifying the assumptions of a standard case of a detailed campaign model. CAPE would be involved to some extent in details, but in connection with capability requirements and risk assessment, not creating data for big models. The Joint Staff would be expected to be a stronger advocate for innovation and a more critical reviewer of service requests. USD(AT&L) would advocate for innovative and sometimes disruptive options that could affect the size and character of the force. These are all nontrivial changes essential to ensuring that ASSP is a meaningful advancement on the preexisting SSA. Some of the important changes will not occur without strong influence from the secretary (hence, some readers will be skeptical).

Analytic Methods

The revised activity described above is impossible without a change in analytic methods. ASSP requires an appropriate mix of methods and tools *and* the means by which to integrate across types of knowledge and levels of resolution. Such integration has precedent but is rare, intellectually difficult, and easy to do poorly. Further theoretical advances and related tools are needed. Nonetheless, much can be done now by having the right people assigned to the effort so that technical, tactical, operational, and strategic-level knowledge is mutually informed.

The new mix of methods should include campaign- and mission-level analysis approached at various levels of detail (multiresolution modeling). Initial priority should be on finding or building relatively low-resolution models that are suitably parameterized.

Staffing

Staffing should be approached by asking what mix of talents and experiences is needed and what mix is appropriate for work within the government, in federally funded research and development centers, or industry. As in the past, the best talent may have disciplinary backgrounds in the physical sciences, engineering, mathematics, economics, and other relevant social sciences. Hiring standards should not associate the title *analyst* with a particular discipline. One objective in rebalancing should be to attract staff capable of higher-level modeling and analysis, crosscutting, and integration.

Research

Solid analysis requires a basis in reality. The new ASSP activity should have a research component to ensure a continuing stream of up-to-date knowledge and capability, as well as an active effort to obtain deep knowledge about the most-crucial subjects (e.g., counterterrorism, survivability of small high-technology ground-force units, robotics, and cyber war). Elements should include the following:

- **Research funds**: Draw knowledge from such diverse sources as history, operational experience, field experiments, operational data, experience of foreign militaries, and the social sciences and then integrate knowledge.

- **Investment in new qualitative and quantitative analytic methods and tools, including human gaming**: Put priority on relatively simple capabilities that deal well with uncertainty and make key assumptions and their significance clear, on methods that exploit social science knowledge where relevant, and on methods that *integrate* across such techniques as human wargaming and modeling. Support work at multiple levels of resolution and require *modularity* to enable sharing, competition, and composition in future campaign models. DoD should *not* contemplate a new, complex, and monolithic simulation.
- **Better leveraging of professional societies**: Staffs benefit greatly from professional activities, and DoD should encourage participation and request the professional societies to have more in-depth meetings and written papers. Often, DoD should commission follow-up activities to generate, for example, tightly constructed definitions disambiguating broad guidance and reconciling different services' understanding of terms.

Caveat: Beware of Halfway Measures

If this report's recommendations are accepted nominally but implementation fails to achieve changes in culture, methods, and staff mix, the result might be worse than the present. ASSP might get in the way of the analysis and decision-aiding that now goes on from non-SSA activities in CAPE, OUSDP, USD(AT&L), and the Joint Staff. Merely revitalizing SSA by restoring past capabilities and methods would be recreating a system already known to be inadequate.

Next Steps

This report says relatively little about implementation, which should depend on many other considerations, such as the preferences of the

Secretary of Defense. A new administration will be taking office in 2017, so preferences could change. A good deal of groundwork should be laid in the meantime.

- DoD should commission a background report to describe in more detail what would constitute a coherent approach to joint analysis that includes low- and medium-resolution modeling and human wargaming, integration of knowledge, and supportive research. This would reflect an assessment of which elements are already in place, which are available but not in place, and which would need development. The report would include definitions, examples, and technical discussion.
- DoD should commission a campaign of briefings and workshops to acquaint the several relevant communities (analysts, decision-makers, etc.) with the ideas proposed and to invite discussion and debate. These briefings and workshops should include outreach to the larger interested community beyond DoD.
- DoD should identify *alternative* implementation strategies, including some human gaming of strategies to better understand the potential consequences, both good and bad. As part of this, DoD should identify sensible time-phasing. It will not be possible to implement all the suggestions of this report in a short period (e.g., establishing multiresolution families of models). A two-year plan would make sense, with important changes occurring early and others requiring more time.

In contrast with the current report, which was mandated as an independent study, this groundwork-laying should involve those currently involved in DoD analysis and the current SSA. The next study will need support and direction from the secretary or deputy secretary.

Acknowledgments

Many individuals provided information and views contributing to this report. I sought to hear the full range of perspectives and issues by meeting with many present and past senior officials, senior officers, and working analysts. Although not responsible for anything in this report, those with whom I met in face-to-face discussions included Chris Skaluba, Beth Cordray, COL Janice King, R. James Mitre, Chris Dougherty, and Jacob Heim of the Office of the Under Secretary of Defense for Policy (OUSDP); Jamie Morin, LtGen Robert Schmidle, Michael Payne, and Elaine Simmons of Cost Assessment and Program Evaluation (CAPE); Tom Carney and James Stevens of the Joint Staff; Nancy Spruill and Mike Knollman of the Office of the Undersecretary of Defense for Acquisitions, Technology and Logistics (USD[AT&L]); Lynne Baldrighi and Mike Cusack of Air Force A-9; Virginia "Robbin" Beale of Navy N81; Daniel Klippstein of Army (Plans and Policy) and Steven Stoddard of the Center for Army Analysis; George Akst and Michael Bailey of the Marines Corps Combat Development Command; Daniel Chiu of the Atlantic Council (previously of OUSDP); Jim Bexfield, consultant, previously of CAPE; and RAND colleagues David Ochmanek, Joel Predd, Yuna Wong, and Igor Mikolic-Torreira, all of whom served previously in OUSDP or CAPE. The report also draws on earlier discussions or interviews by me or RAND colleagues with Vice Admiral Stanley Szemborski, previously of CAPE; Robert Solly of the UK Ministry of Defence; Douglas Feith and Michelle Flournoy, both previously Under Secretaries of defense for Policy; and Ryan Henry, previously Principal Deputy Under Secretary for Policy.

Some of my recommendations draw on earlier research for two under secretaries for USD(AT&L), Michael Wynne and Kenneth Krieg. Stuart Johnson and James Bexfield provided very constructive reviews of the draft manuscript.

CHAPTER ONE

Introduction

Objectives and Background

The objective of this report is to provide an "independent assessment of the joint analytic capabilities of the Department of Defense [DoD] to support strategy, plans, and force development and their link to resource decisions," as called for in the National Defense Authorization Act for Fiscal Year 2015 (U.S. Congress, 2014; see Appendix A).

The report came about because Congress became aware of several concerns relating to DoD's joint analysis. These concerns primarily involved the support for strategic analysis (SSA) activity, which is the main focus of this report. In 2011, the director of the Office of the Secretary of Defense's (OSD's) Cost Assessment and Program Evaluation (CAPE) disbanded campaign-modeling and reduced CAPE's participation in the SSA activity, weakening SSA significantly. The legislation asked for recommendations about capability improvements and related changes of process and organization.

Information-Gathering

The first research phase involved gathering information over several months by in-person interviewing of present and past DoD officials and working analysts. I collected concerns, perspectives, and suggestions on a not-for-attribution basis to encourage candor.[1] I also examined selected interim and final written materials to gain an updated understanding of SSA and its products.[2]

A major event affecting the views of all of those interviewed was the decision by the director of CAPE in 2011 to cease developing SSA's *analytic baselines* and to disband a sizable group charged with campaign-modeling.[3] The decision was controversial and had ramifications that persist to this day and motivated the congressional request for this report.

The concerns expressed by interviewees fell into two groups: those seeing great value in various aspects of SSA, including the campaign-modeling and analysis for which capability is seen to have plummeted, and those giving low marks to SSA processes and some products. The concerns are recounted here as received: During the interview phase, I was listening rather than assessing the correctness or fairness of comments.

A Weakened Analytic Base for Valuable Work

A strongly felt concern was that because of CAPE's dissolution of campaign-modeling capability in 2011, the 2014 Quadrennial Defense Review (QDR) did not have sufficient analytic grounding for its announced strategy and programs. More generally, most of the interviewees agreed that OSD and the Joint Staff no longer have much internal ability to conduct detailed walk-throughs of joint operations and combat-in-planning scenarios at the campaign level.[4] This was said to be true with respect to current, midterm, and longer-term planning. The problem is not just that campaign-modeling is no longer supported but that the related expertise has greatly diminished. Some used the phrase *the baby was thrown out with the bathwater.*

Interviewees also said that the military services no longer have the common databases and baseline cases needed to ensure that, where appropriate, analyses and program development can be based on the same joint data and assumptions. This means that the different services' analyses may be unnecessarily confusing to OSD and Joint Staff leadership. Avoiding that confusion was an important objective in creating SSA and the earlier analytic agenda process. Also, a given service's analysis may be less accurate than desired in representing joint operations and the contributions of its sister services. This is troubling to the individual service chiefs, who intend to be accurate.

Questionable Value to Policymakers

The second class of concerns explained the decision by CAPE to dissolve campaign-modeling capabilities. To a number of senior officials at that time, SSA activity appeared to be neither credible nor useful, particularly when associated with large and complex campaign models. The more that officials learned about the models, they said, the less weight they gave the models in their thinking. Such models generate very different outcomes depending on the many knob settings (assumptions about input variables), which are too numerous and buried to be easily understood.[5] Officials also expressed unhappiness about responsiveness and agility, sometimes mentioning that building scenarios and baseline analyses can take 18 months and that the analysts had been unable to quickly answer "what if" questions when that required changes of important assumptions. Briefers reportedly sometimes responded to questions from senior staff or policymakers with, "Well, that is what the model said" (a cardinal sin in the minds of good analysts). In some cases, senior officials seeking details during spot checks perceived some of the underlying data and assumptions to be noncredible.

To others, the analyses did not contain the information necessary for making strategic-level trade-offs. Many officials in OSD, the Joint Staff, and service staffs criticized the narrowness of the analyses, which did not address the multidimensional uncertainty with which policymakers are afflicted (e.g., what conflicts might arise, in what circumstances, and with adversaries using which strategies and capabilities). This lament was consistent with the independent critiques since the mid-1990s that motivated the Secretary of Defense to refocus defense planning on ensuring broad future capabilities rather than capabilities only for narrow, special cases.[6]

Several interviewees expressed a different concern: The data provided to SSA by services reflect programs of record and traditional concepts of operation (CONOPS), which are not forward-looking or creative. Even if the service in question is studying advanced concepts that it tentatively plans to adopt, these concepts are not reflected in the SSA data (or the service program objective memoranda). Although interviewees understood the services' reluctance to assume success of advanced concepts, they noted that national force planning today demands advanced concepts because advances in the capabilities of potential adversaries have seriously eroded the advantages previously enjoyed by the United States.[7] Further, using traditional concepts may amount to inflating requirements. As a variant of the same point, it was said repeatedly that service inputs to SSA are based on buried assumptions with no transparency. Other interviewees expressed caution about this—pointing out that the services have reasons for conservatism: Often, neither future nor current conflicts play out as envisioned. Also, changes of capability and doctrine often do not develop as hoped (something probably evident to all readers, given U.S. military history since 2001). Nonetheless, the data provided to SSA are problematic. What good are SSA's common databases if senior OSD officials do not trust them or consider them appropriate?

Because of such judgments, these officials did not see the disbanding of the campaign-analysis activity as a significant loss. They saw the action as well justified, especially given the austere budget environment.

Summary of Concerns

The concerns of the two groups of interviewees, then, were seemingly contradictory. Somewhat more agreement existed about the value of SSA's collecting and sharing joint data than about its model-based analytic-baseline work. SSA and its predecessor, the analytic agenda, were warmly credited with having promoted jointness by ensuring that officers in the services were better acquainted with their sibling services' capabilities, doctrines, and perspectives. Joint communication is quite good in comparison with years ago. That has such benefits as allowing analysts to quickly support combatant commanders on current problems or to work effectively in special joint studies as they arise. The sharing of joint data facilitates analysis across all of DoD.

That said, even some of those broadly supporting SSA believed that the detailed analytic baselines prepared in earlier years had led to trouble: The services overfocused on the baseline cases. Many interviewees also felt that the level of detail in the SSA baselines had been excessive. That detail went well beyond specifying objectives and broad strategy—a proper function of the Office of the Under Secretary of Defense for Policy (OUSDP)—and instead specified scenarios and

CONOPS. Further, these interviewees believed that focusing analysis on verifying the executability of a particular detailed scenario and CONOPS discouraged seeking the broader capability analysis needed for planning under uncertainty. This was seen as a serious departure from DoD's intended emphasis on capabilities.

Table 1.1 summarizes primary findings from the interviewing phase of research. This report as a whole aims to illuminate issues and resolve some of the disagreements noted above and in Table 1.1.

First-Phase Conclusions and the Structure of the Report

It was evident from the first phase of research and subsequent review of documents and relevant literature that the problems with DoD's

Table 1.1
Summary of Interview Information

Issue	Views
SSA's value for promoting jointness	Consensus that value was high
The value of SSA's joint data	Consensus that value was high
The value of SSA's detailed analytic baselines	Strong disagreement about value being negative, low, or high
SSA's responsiveness, agility, and understandability	Consensus that problems exist and that senior leaders complain about them; disagreements about seriousness and inevitability
SSA's dealing with uncertainties and supporting capability planning	Consensus that weaknesses exist but disagreements about seriousness
Program creativity within the planning, programming, budgeting, and executing (PPBE) process	Consensus that there is a lack of creativity but with disagreements about resolvability
Logrolling by services and Joint Staff (insufficient debate about service programs or discussion of emerging capabilities because participants accept each other's inputs)	Disagreement between OSD and the military components
The purpose of SSA	Disagreements, except with respect to facilitating sharing of some joint data

capabilities for joint analysis are substantial, that strong differences of opinion exist, and that the root problems run rather deep.

Rather than making on-the-margin tweaks to SSA, DoD should make fundamental revisions to the overall planning construct to which SSA contributes. Doing so will lead naturally to changes in functions, organization, process, methods and tools, and staffing.

The remaining structure of the report is as follows: Chapter Two goes back to fundamentals by asking what the functions of joint analysis should be and what kind of infrastructure of capabilities is needed to perform the functions. Chapter Three then subjectively assesses current SSA using the structure from Chapter Two. The chapter goes on to diagnose problems and identify core assumptions and beliefs that need to be revisited. It also notes enduring tensions that make DoD's joint analysis difficult. Chapter Four is more prescriptive with respect to basic concepts, organization and process, methods and tools, and staffing. Chapter Five provides brief recommendations.

Some Comments on Sources

The first research phase of research involved more than 30 interviews with current or past civilian and military officials, as well as with working analysts. The subsequent phase included selective review of written materials, primarily from 1992 onward, but with examples from the 1970s and 1980s.[8] Some materials were formal U.S. government reports;[9] others were peer-reviewed publications by the National Academies of Sciences and Engineering or federally funded research and development centers (FFRDCs);[10] still others were briefings given at conferences or elsewhere.[11] A few of the materials describe practices of and lessons learned by allied defense ministries, particularly in the United Kingdom, Canada, and Australia.[12] It also looked briefly at some of the classic defense-planning literature dating back to the 1960s.[13] Not surprisingly, some of today's issues were visible then.

CHAPTER TWO

Functions of Joint Analysis and Attributes of Related Infrastructure

This chapter establishes a framework for assessing capabilities for joint analysis. The first section discusses functions to be accomplished; the second section discusses the attributes of infrastructure needed, particularly the part of infrastructure to which SSA relates.

Functions

Strategic planning is an organization's process of considering objectives, formulating strategy, and making decisions about how to allocate resources. It might also include control mechanisms for monitoring, feedback, and adaptation. Strategic planning and strategic analysis occur at multiple levels and in a distributed manner. The Secretary of Defense does strategic planning but so do others within DoD, such as the service chiefs, the Chairman of the Joint Chiefs of Staff, and Under Secretaries of Defense.

Analysis could serve a number of generic functions, such as (1) directly aiding decisionmakers, (2) broadly supporting decisionmakers by providing expertise and undergirding higher-level decision-aiding, (3) supporting execution of decisions, and (4) ensuring that operation planning (e.g., war planning) adheres to strategy and policy. These functions are discussed in the following subsections.[1]

Decision-Aiding

In this context, the word *aid* conveys a sense of direct and personal assistance, rather than just broad indirect support. Ideally, decision-aiding for high officials should[2]

1. frame issues in terms of objectives, criteria, and goals
2. develop and present options
3. comprehensibly compare the options, with pros and cons or by using parametrics for trade-offs and cost-effectiveness analysis
4. provide the real-time ability to "zoom" or "drill down" into detail as necessary to explain particular high-level results or the effect of assumptions[3]
5. discuss, across categories, how to allocate resources so as to balance portfolios (e.g., to decide on *mixes* of capabilities and *mixes* of ways to achieve capabilities, so as to appropriately address all objectives within a budget).

The first four are necessary to address broad strategic issues and for a myriad of component issues. Tensions and conflicts occur across these items, so the fifth one is essential. Decision-aiding is needed by the secretary, chairman, under secretaries, and service chiefs.

Such aiding could be accomplished in personal discussions, a short memorandum or briefing, or a sizable study. In any case, the material should summarize from a much larger body of knowledge and understanding of disputes. Such high-level analysis is, then, in part a matter of asking the right questions, framing the issues, identifying criteria and goals, and providing insights—which is distinct from finding the optimum solution to a well-posed math problem. All such aiding must be comprehensible, credible, and suitable for iteration with decisionmakers. Such decision-aiding is usually the responsibility of elite analytic groups working personally for the decisionmaker.[4] It is not a natural function for committees.

Broad Underlying Support of Strategic Decisionmaking

Broad support for decisionmaking includes ensuring that knowledge is available when needed on diverse subjects. Broad support includes providing a deeper body of knowledge and information. After all, high-level decision-aiding depends on many simplifications and judgments. These should be informed by broader and deeper knowledge, often based on research, analysis, and empirical data. Table 2.1 illustrates the kinds of DoD issues on which such deeper knowledge has been needed in the past. Each of the questions led to its own joint study drawing on SSA information. Ideally, the products can flow upward, as suggested in Figure 2.1. The figure anticipates that decisionmakers will use their own strategic analysts to produce the needed products. Ideally, however, these products will be informed by deeper analysis. As an example, while preparing an analysis on strategic mobility for decisionmakers, it is of little use to know all the data that went into a detailed study showing that a particular deployment could be accomplished with the projected level of airlift. Rather, the analysis should show parametrics from such deeper analysis, parametrics indicating what kinds of deployments could and could not be supported. The parametrics should be consistent with (although abstracted from) more-detailed work.

Table 2.1
Illustrative Questions in Need of Deep Analysis

Question
What capabilities does the United States need to find and neutralize weapons of mass destruction?
What capabilities does the United States need to execute its strategy for the war on terror?
What is the best mix of tactical air forces?
What is the best composition (number, mix, and characteristics) of the future tanker fleet?
What capabilities does DoD need to support civil authorities when responding to a domestic disaster?
Are U.S. projected future forces sufficient to meet the defense strategy?

SOURCE: Stevens, 2008.

Figure 2.1
Relationship Between Broad Support and Decision-Aiding

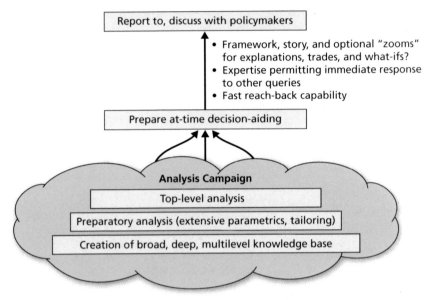

SOURCE: Davis, 2014, Figure 3.1.
RAND *RR1469-2.1*

Where such deeper knowledge is needed, supporting analysis should generally

1. reflect a full understanding of the challenges, conceiving the system broadly
2. consider broad and creative options and alternative ways to frame issues
3. provide clear and rigorous definitions, distinctions, assumptions, metrics, and goals
4. provide *multiresolution* analysis of uncertainties and disagreements (e.g., parametric trade-offs at different levels of detail, as needed)
5. arrange for research to better understand the phenomena involved in the challenges and obtain data

6. winnow down by suggesting simplified specialized frames, measure hierarchies, and best-of-breed options for each option type to be sent up the line and perhaps used in high-level decision-aiding

7. for all of these items, focus on how to achieve flexibility, adaptiveness, and robustness (FARness)—rather than solutions fine-tuned to precise but fragile assumptions about the future.[5]

The seventh item has not been recognized adequately in traditional supporting analysis. This item defines a new demand that policymakers should place on analysis and a new professional responsibility for analysts that goes well beyond identifying assumptions, as good analysts already do.[6]

Supporting Execution

A sometimes-neglected element of analysis for strategic planning is supporting the execution of decisions. This includes the following:

1. Interpret and tighten the decision and then propagate more-detailed guidance, including measures, metrics, and goals.[7] Without this, guidance may be difficult for components to follow and too vague to force desired changes.[8]

2. Monitor performance and outcome from related initiatives and supporting lower-level adaptations: Are programs funded and executed as intended (inputs); are they proceeding as intended; are their outputs as intended?

3. Flag variations from the plan that imply the need for review and new top-level decisions. Examples might involve failure assumptions critical to the original decision: changes in the world, technical failures, intolerable cost growth, or discovery of fatal flaws in the CONOPS.

Getting the metrics right is difficult. Organizations often come up with metrics that seem appropriate superficially but are actually useless or worse because they create perverse incentives. Sound metrics need to be derived *analytically* from an understanding of operations. In many cases, these metrics turn out to be natural parameters of a good model of the operation and also to fall into natural levels of hierarchical detail.[9] Using higher-level parameters that are indifferent to details of how an operation is accomplished can avoid some of the pernicious effects of metrics that presuppose such details.

Ensuring Adherence to Strategy and Policy in Operational Planning

Another function of joint analysis is to ensure that the intentions of DoD's strategy and policy are properly reflected in operational planning, such as war planning by combatant commanders, and in the guidance that they receive through the Joint Staff (e.g., the Joint Strategic Capabilities Plan). Most aspects of this are classified, but the intentions are indicated by definitions and legislation.[10] This report does not address such matters.

Functions of SSA

Against this background of generic functions, the SSA activity was intended primarily to accomplish two purposes as specified by a DoD directive (DoD Directive 8260.05, 2011): (1) Support deliberations by DoD senior leadership on strategy and PPBE process matters, including force-sizing, force-shaping, and force-capability development, and (2) provide a starting point for studies that support development and implementation of defense strategy and policy and support DoD's PPBE process.

The five functions of analysis are fulfilled in different ways. Decision-aiding is primarily the responsibility of staffs working directly for the decisionmaker. SSA's role has been largely to provide broad underlying support in the form of expertise, joint data, and joint analysis baselines (the shaded area in Table 2.2). SSA has not played much of a role in supporting program execution; for the Secretary of Defense, that function is largely served by CAPE, assisted by the Joint Staff and by service staffs. SSA is also only playing a limited role in ensuring adherence of war plans to strategy, although it could do more.

A cooperative, consensus-seeking activity such as SSA cannot reasonably be very useful in directly aiding decisionmaking by the Secretary of Defense (the activity can be more useful for aiding service

Table 2.2
The Support for Strategic Analysis Role in Joint Analysis

| | Decision-Aiding | Broad Support | | Support of Execution | Ensuring Adherence to Strategy |
		Expertise	Undergirding Higher-Level Analysis		
Current role of SSA	•	••••	••		
Possible future role of an SSA-like activity	•	••••	••••	•	•

NOTE: More bullets in a cell indicates a greater emphasis on the column's function.

chiefs) (A. Barber, 2014a, 2014b), but the activity can provide broad support in the form of expertise and the undergirding of higher-level analysis. A future SSA-like activity could do much more in this respect, with its results flowing coherently into modern, uncertainty-sensitive, higher-level analysis for decision-aiding (Figure 2.1). The activity could also play more of a role in supporting execution and ensuring adherence to strategy. Any such changes, if desired, would require significant changes in expectations, organization, tasking, process, methods, and staffing.

Attributes of Infrastructure

After discussing functions of joint analysis, it is useful to ask what is needed to have a good *infrastructure* for the broad range of analytic activities. In this view, the outputs of the infrastructure are the quality and responsiveness of the functions. But what attributes constitute a good infrastructure, one that can perform these functions? And, in particular, what should be the attributes of SSA when viewed as infrastructure? A reasonable set of attributes to consider is

1. diversity of suppliers
2. healthiness of relationships among suppliers and with clients

3. sufficiency of human and intellectual capital of the suppliers (including methods)
4. adequate processes to manage and sustain the infrastructure.

For DoD, the infrastructure for joint analysis must include participants from all the relevant DoD components, with participants understanding their joint contexts and narrower ones, in addition to being able to coordinate and even collaborate well.

The components of DoD are competing for resources, which strongly affects individual incentives with DoD.[11] A good infrastructure for DoD should make it possible to manage such competitions fairly (ensuring a level playing field when comparing service programs), wisely, and effectively at reasonable cost. That implies requirements for openness, sharing, and common basic data. In addition, it requires management attention and resources to keep the infrastructure diverse, vibrant, learning, and keeping up with and even anticipating new challenges. The next chapter evaluates SSA by these criteria and those given earlier in this chapter.

CHAPTER THREE

Evaluation and Diagnosis

Base Evaluation

Evaluation of Functions

Congress asked for an assessment of DoD's current capabilities for joint analysis (see Appendix A). Resources did not permit an exhaustive study. What follows are my subjective, though informed, assessments using the structure laid out in Chapter Two, the considerable interview information gathered during the study, and previous experience and research (see Davis, 2014). The assessments appear approximately correct, and some scores are given as ranges. Significantly, a given reader might disagree with the *precise* subjective scores assigned while still being convinced of the shortcomings identified in the subsequent discussion.

Table 3.1 defines the qualitative, subjective assessments on a 1–9 scale, with 1 being very poor and 9 being very high.

This scaling is probably intuitive to readers who have been consumers of analysis. Some analysis lacks credibility or is irrelevant to decisions, or both (score of 1). Perhaps the problem is low-quality analysts, but more likely it is because the issues are not well understood. A good deal of analysis merits a medium score of 5. Here the analysis is credible as far as it goes, relevant, and responsive. It also provides some significant insights. It is limited, however, by narrowness and a somewhat pedestrian flavor—answering specific questions but not generalizing beyond the specifics. Also, this analysis might not boil issues down or cleanly frame issues. At the high end (a score of 9), analysis is not only very good, relevant, and responsive but goes beyond that to

Table 3.1 Definitions of Scale Values in Subjective Assessments

Score	Meaning
1	Poor, unresponsive information; little insight; little expertise; no useful abstraction or framing
3	A mix of 1 and 5 (i.e., examples of each)
5	Good, semiresponsive information; somewhat narrow expertise; some insight; not much uncertainty analysis or trade-offs, imagination, anticipation, framing, or useful abstraction
7	A mix of 5 and 9 *or* consistent in-between performance
9	Very good, highly relevant information; broad expertise; imagination, uncertainty analysis, anticipation, framing, and useful abstraction with supporting analysis

sharpen and expand the questions, anticipate larger issues, and provide more-general insights (e.g., indicating under what circumstances results would and would not be favorable). Finally, it suggests simplifications.

All of this begs the question of whether we are measuring analysis relative to what was asked or relative to what *might* have been asked. Perhaps policymakers should pose different and more-thoughtful challenges for SSA. With that in mind, Table 3.2 provides this report's assessments for the current standard (i.e., for the tasking as it has existed and been interpreted) and for an elevated standard. The later assessments are based on my subjective sense of how well the 2011 SSA could have done before the loss of its campaign-modeling had questions been more demanding.

Decision-Aiding

The second column of Table 3.2 shows the assessment of SSA for the decision-aiding function. This has *not* been a core SSA function, so it is not surprising that SSA has not contributed significantly to directly aiding the secretary or under secretaries. However, those interviewed noted with examples that service chiefs have sometimes been directly aided. Thus, even if the overall assessment is 2, the range is perhaps from 1 to 5. The score might have been somewhat better had different questions been asked (the elevated-standard row).

Table 3.2
Top-Level Subjective Evaluation of Current Support for Strategic Analysis as a Function of Standard Used

| | Score for Aiding High-Level Decisionmaking | Score for Broadly Supporting High-Level Decisionmaking (current role of SSA) | | | | Score for Supporting Execution | Score for Supporting Adherence to Strategy and Policy |
| | | Expertise and Advice | | Undergird High-Level Analysis | | | |
		2011	2016	2011	2016		
Current standard	2 (1–5)	5 (4–7)	4 (3–5)	3 (1–5)	2 (1–3)	N/A	1–3
Elevated standard	3 (1–6)	6	4	5	3	Selective possibilities: definitions, metrics, goals, and risks	Some role possible
Reason for score	Mismatch with Secretary of Defense needs (better match with needs of chiefs)	Problems with level of analysis, staffing, and mind-sets	Problems with level of analysis, staffing, mind-sets, and reduced capabilities since 2011	Problems with level of analysis, staffing, and mind-sets	Problems with level of analysis, staffing, mind-sets, and reduced capabilities	Problems with level of analysis, staffing, mind-sets, and reduced capabilities	Levels; methods; and problems with level of analysis, staffing, mind-sets, and reduced capabilities

NOTES: The scale is 1 (very poor) to 9 (excellent). Ranges are in parentheses. N/A = not applicable.

Supporting the Execution and Review of Current War Plans

Skipping temporarily to the last two columns of Table 3.2, the assessment is that SSA has not been expected to support the execution of the defense program. As seen in the elevated-standard row, it would seem logical that SSA could have done more, primarily to sharpen definitions, suggest metrics, provide analysis that suggests goals for the metrics, and characterize execution risks. That would be very difficult now because of the reduced campaign-modeling capability.[1] As for supporting OSD's review of current war plans (last column, on supporting adherence to strategy and policy), the Deputy Assistant Secretary of Defense for Plans called on SSA-related expertise to some extent through 2011. With the loss of SSA's campaign analysis, however, that expertise is no longer available.

Expertise and Undergirding Higher-Level Analysis

The shaded columns in the middle of Table 3.2 show assessments for the core functions of SSA. These are shown separately for the period through 2011, when SSA had extensive campaign-modeling, and the subsequent (current) period, when SSA has not had such modeling. The assessment is that, through 2011, SSA provided significant expertise (those doing campaign-modeling were professional and respected).[2] Nonetheless, SSA was judged as falling short in terms of its work undergirding the more-important aspects of high-level analysis. Those more directly in the joint decision-aiding business did not see much benefit to SSA's efforts and felt that the analytic baselines were getting in the way by underplaying uncertainty and the search for fulsome capabilities. Also, the efforts overfocused on the campaign level, rather than the mission level where many problems reside. Some saw SSA as an unproductive overhead activity.

Table 3.3 explains the evaluations of the undergirding-strategic-analysis function using criteria from Chapter Two. SSA has done *some* framing, has arguably provided *some* options (usually as provided by the services), and had done at least a bit (score of 3) in comparing options. It has not, however, done much uncertainty analysis or provided layered explanations. Because point cases are of little interest, the score assigned is 3. SSA could have been asked to do more (the ele-

Table 3.3
Factors in the Assessment of the Support for Strategic Analysis Function of Undergirding High-Level Analysis

	Score on Framing	Score on Providing Options	Score on Comparing Options Comprehensibly and Credibly, with Uncertainty Analysis and Layered Explanations (zoom)	Score on Allocating Resources	Score on Undergirding High-Level Analysis (2011)
Current standard	5 (3–7)	4	3	N/A	3
Elevated standard	5 (3–7)	5	5	N/A	5

NOTES: The scale is 1 (very poor) to 9 (excellent). Ranges are in parentheses. N/A = not applicable.

vated-standard row), and the results would probably have been somewhat better. SSA probably could not reasonably have been asked to help in allocating resources. SSA depends on cross-component cooperation, sharing, and sometimes on consensus. Such an activity is seldom good at resource-allocation analysis, which is why organizations such as CAPE exist. Overall, then, SSA did not fare well in the assessment (see the last column of Table 3.3).

Evaluation of SSA Infrastructure

Table 3.4 is the assessment of SSA as infrastructure. The shaded columns use the attributes described in Chapter Two. Overall, the evaluation is moderate (5) with a sizable range (4–7).

Senior leaders see relatively little of what goes on at the SSA level. They may not easily appreciate such infrastructure. I believe, however, based on past experience and the testimony of those interviewed for this study, that the SSA infrastructure has been very important in DoD's nurturing of jointness. Officers and analysts who would otherwise be stovepiped in their services or OSD offices mingle and work together routinely and well—learning about each other's capabilities, doctrine, special features, and idiosyncrasies (including those of lan-

Table 3.4
Evaluation of Support for Strategic Analysis as Infrastructure

Products of SSA Infrastructure		Attributes of Infrastructure				Assessment
Score on Quality	Score on Responsive-ness	Score on Diversity of Suppliers	Score on Relationships Among Suppliers and Clients	Score on Human and Intellectual Capital	Score on Infra-structure Oversight	Score on Evaluation of Infra-structure Attributes
5 (3–7)	3	3	7 (4–8)	5 (3–7)	6 (3–7)	5 (4–7)
Some high-quality products; narrow	Poor[a]	Weak on technology push and advanced concepts	Mostly good given enduring tensions; inadequate confrontation	Good but narrow		Limited by authorities, complexity, consensus, resources; includes some research

[a] SSA-associated analysts have often been responsive to their immediate leaders in CAPE, OUSDP, the Joint Staff, and the services. The analysts' responsiveness has often benefited from knowledge and data gained as part of the SSA activity.

guage). These officers and analysts learn the trade of joint analysis by using common methods and models. This is akin to having common textbooks. This pays off when Joint Staff analysts are called on to help combatant commander staffs in theaters of war or when DoD has an important joint study. SSA, however, has had weak connections with the worlds of technology and innovation (e.g., the focus of the recently created Strategic Capabilities Office, which reports to the Deputy Secretary of Defense).

Summary Evaluation and Diagnosis

In summary, this report's assessment of SSA is not very positive. Further, SSA's capabilities and value have dropped significantly because of the loss of SSA's capability for campaign-modeling. In my view, however (consistent with many of the inputs received),

- restoring past campaign-modeling capability would not resolve the largest problems

- *major* changes are needed and feasible, as discussed in Chapter Four
- the basic problem is that current SSA has severe limitations of methods, tools, staffing, and mind-set.

This assessment is not a criticism of SSA's efforts. SSA had, by 2010–2011, become effective at what it had been asked to do and was on a path toward doing even better.[3] SSA's history includes numerous bright spots, including efforts to address difficult problems relating to counterterrorism, counterinsurgency, and intervention.[4] SSA has much to be proud of. That said, for a revitalized SSA to achieve the elevated standards that it should be given to better serve policymakers, major changes will be needed in concepts, process, organization, and staffing. That will require some rethinking by senior leaders as they consider a redesign.

Comparison with an Earlier Assessment of DoD Analytic Culture

Many issues are long-standing. Table 3.5 is taken from DoD modeling and simulation (M&S) conducted in 1996 (Defense Science Board, 1996a). Some of the shortcomings that the Defense Science Board noted of DoD analysis culture have been mitigated since then. DoD now considers many more scenario types,[5] and it has more joint data to support analysis. It has also taken a closer look at such complex phenomena as counterterrorism and counterinsurgency. Other shortcomings, however, are the same as they were 20 years ago: complex bureaucratic processes, suppression of deeper forms of uncertainty, and "mechanical" models (i.e., models that are longer on computation than on providing insight).

The Defense Science Board study also suggested contributions to infrastructure, as indicated in Table 3.6. The study suggested a dedicated research organization established by the Director for Defense Research and Engineering, what is now roughly the Assistant Secretary of Defense for Research and Engineering, to study information dominance, long-range precision strike, and operational concepts. This research was to be tied to the warfighting community and would be expected to solve new problems. The Defense Science Board also

Table 3.5
Changes Needed in the DoD Analysis Culture (from Defense Science Board Study)

What Is	What Should Be
Closed processes	Open processes
Bureaucratic review	Peer review
Accredited analysis	Competitive analysis
Model orientation	Subject-matter orientation
Mechanical	Meaningful
Data-poor	Data-rich
Rigid approvals	Learning and adaptation
Stable algorithms	Unstable phenomena
Suppressed uncertainty	Illuminated uncertainty
Suppressed risk	Illuminated risk
Narrow, Cold War style	Oriented to present and future
Few accredited scenarios	Many political-military scenarios
Point assumptions	Exploratory analysis

SOURCE: Adapted from Defense Science Board, 1996a (a report chaired by retired Air Force Major General Jasper Welch).

envisioned a holistic view of M&S within analysis, as indicated in Figure 3.1.

The Defense Science Board's suggestions pertained to DoD joint analysis broadly, not just to what became the work of the later analytic agenda and SSA. SSA has been tied more to PPBE than to acquisitions. Some of the Defense Science Board's recommendations have been pursued under sponsorship of USD(AT&L). USD(AT&L), for example, sponsored the creation of the Joint Advanced Warfighting Program at the Institute for Defense Analyses.

In some respects, it has been desirable for such work to be separated from that of PPBE because of the cultural differences between communities.[6] One consequence, however, has been that SSA has suf-

Table 3.6
A Research Approach Suggested in 1996

Mission Statement	Examples
Identify analytical issues needing improvement derived from both customer concerns and research	Countermeasures, realistic effectiveness calculations, behavioral assumptions at both small-unit and commander level
Collect and analyze empirical data from all sources, both existing and program-generated	History, structured interviews, instrumental training exercises, virtual exercises, field tests
Collect and analyze results from all types of M&S, both existing and program-generated, at all relevant levels of resolution	Encourage integrated hierarchical families of models, including selectable resolution; exploit data from all levels of resolution
Create and maintain an overall intellectual framework by engaging customers, actively guiding the research program, and stimulating both peer review and open debate	Provide strong problem definition and analytic plans to those conducting virtual experiments. Create subject-area forums for in-depth exchange and peer review
Serve the customer community by providing expert advice and advisors, making available new and better analytical modules for widespread use, and evaluating analysis at request of customers	Provide red teaming, definitive effectiveness calculations, and analytically sound modules (while recognizing uncertainty)

NOTE: Retyped and reformatted, but otherwise identical to original (Defense Science Board, 1996a).

fered from inadequate technology push and innovation push.[7] Also, DoD as a whole has nothing similar to the program suggested by Figure 3.1. Some related suggestions appear in Chapter Four.

Comparison with Suggestions from a National Academies Study

As another data point to demonstrate that the problems noted in Table 3.5 and earlier in the report are long-standing, Table 3.7 mentions selected recommendations of a National Academies of Sciences, Engineering, and Medicine study from 2006. The study was supported by the Defense Modeling and Simulation Office[8] through an initiative of CAPE. The panel referred to modeling, simulation, and analysis (MS&A) to emphasize that DoD's focus on M&S was somewhat wrong-headed because it omitted the crucial element of high-quality

Figure 3.1
A Holistic Approach Suggested in 1996

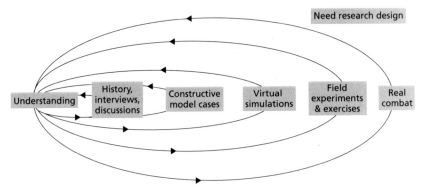

SOURCE: Adapted from Defense Science Board, 1996a.
NOTES: Research includes empirical work and model-based work that vary in levels of resolution, types, and degrees and human and analytical exploration of uncertainty and sensitivity. The intent is to use or generate, and then integrate, all knowledge of a phenomenon.
RAND RR1469-3.1

Table 3.7
Selected Recommendations from 2006

Some Synopsized Recommendations
Modern technology and CONOPS: Ensure that the basic architecture of MS&A systems reflects modern concepts of network-centric warfare.
Broad approach to analysis and methods: Invest in a range of methods, including diverse models, games, field experiments, and other ways to obtain information.
Social-science methods: Devote significant research to social behavioral networks and multi-agent systems because both are promising approaches to the difficult modeling challenges.
Uncertainty: Seek better methods to characterize, quantify, and manage the uncertainty inherent in all aspects of MS&A—including inputs, modeling assumptions, parameters, and options.
Interface with decisionmakers: Strive to understand cognitive styles of decisionmakers and their interaction with different forms of MS&A. DoD should seek better methods to characterize interface.
Broad research on military science: Identify (or create) and charge an organization with responsibility for developing and supporting research and development to improve and update base of military science for combat and noncombat modeling.

SOURCE: Adapted from National Research Council, 2006.

analysis—something that does not follow easily from M&S by itself and is sometimes frustrated by the nature of M&S. By and large, the recommendations have not as yet been heeded.

Diagnosis

Root Problems: How Joint Analysis for Defense-Planning Is Conceived

The roots of SSA's shortcomings can be understood by observing a graphic (Figure 3.2), which simplifies but explains what SSA is supposed to do and in fact did do before dissolution of its campaign-modeling capability (see Appendix B). In 2011, the process began with strategy and was followed by OUSDP-led development of defense planning scenarios expressed as rich essays depicting intentions of strategy and the complex challenges of executing it. The essays included dilemmas and uncertainties; they painted a challenge *space*, not a point. The Joint Staff then developed concepts of operations to execute the strategy for the broad scenario painted by the defense planning scenarios.[9] That was followed by collaborative studies of the campaign-modeling variety to establish enough context to allow specification of the myriad inputs used by campaign models. In this 2011 depiction, this effort was led by CAPE's Simulation Analysis Center (SAC). An analytic baseline was the full package of scenario, CONOPS, context-setting study, and integrated joint data. The analytic baselines were intended to be baselines in the sense of a common point of departure for numerous and substantial variations performed mostly by the services. The baselines themselves demonstrated the feasibility of executing the OUSDP-provided strategy, but identified difficulties in doing so that were itemized in massive documentation (usually by identifying assumptions). Over time (from year to year), insights from one cycle informed the next round, even strategy itself.

By 2010–2011, SSA was largely successful in carrying out this process. I personally reviewed illustrative SSA analytic baselines and was impressed by the quality of the scenarios, the quality and documentation of the CONOPS (more properly referred to as *Multi-Service*

Figure 3.2
Simplified Depiction of the Support for Strategic Analysis Process (Through 2011) for Midterm Planning

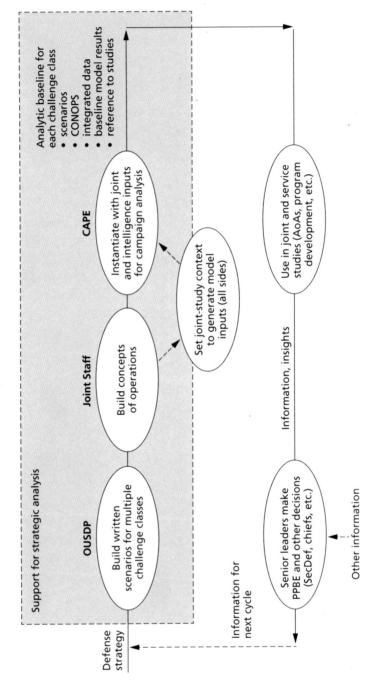

NOTE: PPBE = planning, programming, budgeting, and executing; SecDef = Secretary of Defense; AoA = analysis of alternatives.

RAND RR1469-3.2

Force Deployments), and the valuable body of knowledge represented by the integrated data. By 2010–2011, SSA was doing well at what it had been tasked to do.

As of 2011, CAPE no longer supports campaign-modeling and no longer creates the analytic baselines. SSA continues to distribute older baselines and is attempting to proceed in different ways. In particular, the Joint Staff's J-8 has picked up some of the burden on an ad hoc basis, and the services are attempting to coordinate more directly. CAPE and OUSDP are collaborating to identify and sharpen key issues, many of which are mission-level challenges. The situation is far from satisfying to participants, but work continues despite frustrations.

It may seem, at first glance, that the SSA process in Figure 3.2 was eminently sensible, in which case restoring all the pre-2012 functionality would seem to be important. However, based on results of this study, I concluded otherwise.

Strategy is provided as a given (the left side of Figure 3.2). This has the effect of relegating analysis and support thereof to something like operations research: A problem is presented and analysts are then asked to show how the problem can be solved (developing CONOPS and demonstrating executability). Strategy, however, does not come out of nowhere at the whim of leaders. Rather, it is a *product* of initial analysis in OUSDP. This includes significant background research,[10] as well as informal analysis that reconciles objectives, feasibility, and the resources plausibly obtainable. Should an SSA-like activity not support such analysis? And why should such analysis not be more explicit? Why should SSA *only* provide data rather than do supporting analysis?

Scenarios are shown as flowing from strategy, but should there be explicit analysis supporting the development of scenarios and should an SSA-like activity be part of it?

The step of CONOPS development requires the Joint Staff to address a given scenario, develop an initial CONOPS, and then make countless other assumptions about subsequent operations and how they will turn out. Even if developers make good choices, the result is essentially the sketch of a detailed point scenario. Developers of the CONOPS are unable to address all the challenges posed in the scenarios because developers have to choose which paths to take among those that are possible. Developers can and do build in some clever branching, but there are severe limitations.

The joint data and analytical baselines are even narrower in scope. Further, in practice, they have been developed to feed rather-detailed campaign models, which—as currently used—are not well suited for big-picture thinking and *broad* uncertainty analysis.[11] Also, they are not well suited for addressing many of the softer aspects of operations, including counterinsurgency, counterterrorism, and stabilization.[12] Thus, referring to these model inputs as data fails to convey a sense of what may be involved.

The overall process does not even mention uncertainty analysis, although there is some uncertainty analysis used to construct the analytical baselines within the joint study. As noted, the assumption behind SSA was that the services or special joint studies would use the baselines as points of departure for subsequent broad exploration, but, in practice, such exploration was very limited—to the continued frustration of OUSDP and others. The baselines became end points. Also, the baselines were sometimes interpreted by the services as specifications. This was seriously problematic because executability of baselines may be a *necessary* requirement, but it does not ensure the ability of the programmed force to deal with all the many variants in the original (essay-form) expression of scenario. *No single point scenario can represent the set of necessary and sufficient requirements.*[13] The last point can be expressed differently: The process that has emphasized standardization of scenarios, CONOPS, and input data for complex campaign models has undercut the very intentions of defense secretaries seeking to build capabilities (forces and the technical capabilities they possess) that can do well for diverse situations. This was the idea behind capability-based planning, dating from the mid-1990s and reflected in the 2001 Quadrennial Defense Review (Rumsfeld, 2001).[14]

Mission-Level Versus Campaign-Level Analysis

As a related matter, CAPE had concluded by 2011 that many of the issues it sees as critical must be studied on a mission-by-mission level, considering the many uncertainties facing each issue. It is the possible variations that matter, not the notional result of any single base case. Heavily standardized campaign analysis had gotten in the way.

Those disagreeing with CAPE argue that campaign analysis is logically and practically necessary: They agree that having capabilities is necessary and critical, but a key question is whether the capabilities can be put together to accomplish what is called for in strategy. Can the strategy be executed (a key issue in force-sizing)? How else can that be assessed except by campaign analysis?

The synthesis here is, of course, that joint analysis needs to do both.[15] Doing so requires recognizing the need for different *levels* of analysis and related levels of modeling (multiresolution modeling). Conducting a campaign is to conduct a sequence of missions (some of them at the same time). Success of the campaign requires success in all of the critical missions. Campaign analysis, however, need not go into the details of how the missions are accomplished. Lower-resolution campaign analysis can be (and was, in earlier decades) about ensuring the ability to deploy adequate forces to one or multiple places in time to succeed with strategy. What constitutes an adequate force may be estimated from various mission-level analyses (as well as historical data and doctrine). Although it is common for DoD campaign analyses to have considerable detail (e.g., the war-fighting detail that includes attrition, rates of advance, and other dubious predictions), that is often unnecessary and even distracting for informing strategic decisions.

Good People Can Partially Compensate for Organizational Process Shortcomings

Interestingly, *outcomes* of DoD processes (e.g., essential features of QDRs) are often better than the processes themselves. This is because dedicated people find ways to do what is most needed, despite shortcomings of organization and process. The product can be good even if the process is not always neat. Indeed, currently, the *actual* process is richer and more subtle than graphic depictions: Some of the crucial thinking and analysis are accomplished informally, whether in CAPE, OUSDP, or the Joint Staff. Nonetheless, the remainder of this report assumes that DoD would do better if some of that thinking and analysis were more explicitly recognized and approached with a reasonable degree of rigorous analysis.

Enduring Tensions

Before turning to prescriptions in Chapter Four it is useful to summarize some enduring tensions that make defense analysis, and support of joint analysis within defense analysis, difficult. Many of these tensions appear routinely in strategic planning activities of many organizations.

Standardization

Mature organizations like standardized processes, models, data, and planning scenarios. People typically want to know what is expected of them, what they are agreeing to do, for what they are *responsible*, and for what they will be rewarded. All this may, however, be at odds with what is needed to deal effectively with change, uncertainty, and disagreement. The tension is not unique to government or DoD, as noted in the academic literature.[16]

There is a clear need for common data sets to allow the comparison of analyses using the same assumptions (a long-standing demand by senior leaders who receive multiple briefings that cannot be compared directly). The tendency, however, is to never get around to the uncertainty analysis. The solution is not to avoid baselines but to demand the excursions. A stumbling block has been the methods familiar to DoD's joint analysts, as discussed in Chapter Four.

Uncertainty

Development of traditional defense-analysis methods and models was begun decades ago, when the concepts for dealing with uncertainty were poorly understood and when it was common to sweep them under the rug. This was an embarrassing lapse noted even in the 1960s (Quade and Boucher, 1968; Quade and Carter, 1989, pp. 354–355). One reason was an underlying tension: If one tried to address uncertainty, it was possible to be overwhelmed by complexity and to be paralyzed. Also, early practitioners sought scientific rigor but interpreted that to mean hard quantification and prediction. Today, strong methods exist for dealing effectively with uncertainty without paralysis. Further, modern thinking about science and rigor requires dealing well with "soft" uncertainty, not wishing it away. To be sure, many senior officials seek predictive analysis even when it makes no sense, but such officials as the defense secretaries, Joint Staff chairs, and service chiefs are fully aware that they are planning under deep uncertainty. They have not been well served by analysis that suppresses uncertainty.

Level of Resolution (Level of Detail)

A tension always exists between the desire to look into detail and the desire to see the whole. Similarly, there is tension between seeing the particular and seeing the realm of the possible. Organizations also recognize that "information is power" and that they may benefit by framing issues in detail and then controlling that detailed information.

In fact, knowledge and analysis are needed at different levels of detail. Senior leaders must focus at a high level, but they must rely on those below them to know and deal with the details. An SSA-like activity should provide such details but also the path to the higher-level reasoning. Policymakers want to understand the issues and options, which implies reducing the issues to manageable proportions. Also, as stressed over the years, the most-effective leaders make decisions based on general considerations rather than focusing on finding elegant solutions to very specific but ephemeral problems (Drucker, 2006).

Analysts dependent on computer models are especially prone to losing the forest for the trees. The campaign models of DoD and analogous models used in other domains involve tens of thousands of lines of computer code, with countless inputs that are a mix of hard data, reasonable estimates, and very unreliable assumptions. Better examples could hardly be found of false precision than using such complex campaign models and imagining the results to be predictive, except perhaps broadly. All of this implies a need for multiresolution research and analysis and related crosscutting methods, as will be discussed in Chapter Four.

False Optimization

Many analysts have been trained to seek optimal solutions for problems posed by senior leaders. These analysts are, in a sense, solving a math problem. Senior leaders, however, recognize that this can be counterproductive because of uncertainties. Instead, senior leaders seek strategies that move in the right direction, creating the capabilities for flexibility, adaptiveness, and robustness to deal with realities as they arise. To the extent that the policymakers have utility functions, those sometimes emerge in the making of decisions rather than being something stable known in advance. Moreover, the utilities will change as world events occur. It follows that optimization can be one of the powerful tools useful in the depths of a good analysis organization but not as the centerpiece for resolving most senior-leader issues.[17]

Consensus

Organizations like consensus, as do many senior officials and officers. The search for consensus, however, can be both paralyzing and counterproductive. This is now a well-known problem in DoD, so it need not be elaborated here.[18]

CHAPTER FOUR

Prescriptions

Overview

Based on the previous chapter's background of assessment and diagnosis, this report recommends that DoD replace SSA with a new activity rather than make incremental fixes. This is because many of the problems noted in Chapter Three are of a fundamental nature. In this chapter, subsequent sections deal with (1) reconceiving the functions of joint analysis and support thereof, (2) implications for organization, (3) implications for methods, and (4) implications for staffing.

Some themes that run through the chapter are as follows:

- See strategy as an *output* of analysis and deliberation rather than as a stand-alone input.
- Characterize the degree to which the strategy provides for as much flexibility, adaptiveness and robustness (FARness) as possible within budget constraints. Also, characterize the implications of smaller or larger budgets in such terms, thereby providing more information about risk.

With these outputs in mind, it then becomes necessary to do the following:

- Reconceive the questions asked of analysts and the kinds of information demanded from them.
- Adjust organizations and processes accordingly.

- Broaden and change the character of analytic methods and research on which to base analysis.
- Adjust the mix of staff accordingly.

Reconceiving Functions and Attributes of Joint Analysis

If the joint analytic activity is to support strategic planning, it is appropriate to consider what the domain of such planning is. Figure 4.1 is my version of a diagram often shown by senior officials. When referring to capability-based planning in the context of strategic planning, another characterization sometimes used recognizes several levels of decision:[1]

- Level 1—help senior leaders manage risk across the challenges and determine the best balance of investments across major capability areas (e.g., QDRs)

Figure 4.1
The Domain of Strategic Planning

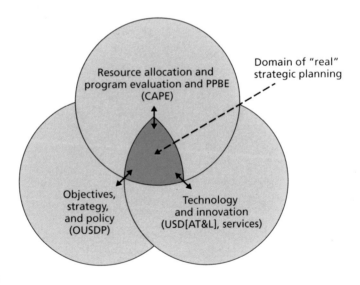

- Level 2—Help determine how best to accomplish missions or joint concepts (e.g., Joint Capabilities Integration Development System studies)
- Level 3—Help decide which systems to buy or stop buying and establish design parameters for new systems (e.g., analyses of alternatives).

In contrast, and as noted in Chapter Three, SSA has had only weak contacts with technology and innovation. And SSA has not done much to support analysis of mission and capability-area issues. Thus, SSA has been supporting only some aspects of strategic planning.

A New Construct and Process

Figure 4.2 shows a revised construct for analysis to assist in mid- and longer-term planning.[2] The lightly and more darkly shaded portions show the new activity, which reflects the report's suggested objectives for planning midterm capabilities.

Against this background, this report recommends that DoD create a new activity to replace SSA. It could be called *analytic support for strategic planning* (ASSP).

1. ASSP should support *initial, interim* decisions on defense strategy and budgets with explicit, understandable, and therefore relatively simple analysis.
2. ASSP should emphasize planning for flexibility, adaptiveness, and robustness (FARness); ASSP should deemphasize detailed analytic baselines. As occurs now, it should develop a list of type scenarios (e.g., defeat a particular rogue state that is attacking an ally, prevail in two simultaneous wars, and cope adequately with even more-complex cases).[3] Then, *for each*, ASSP should develop spanning sets of variations to stress U.S. capabilities in all the dimensions needed. The spanning-set scenarios should constitute necessary and sufficient requirements for the services to meet in their program development.
3. ASSP should include options that incorporate emerging technology and innovative concepts.[4] To assist in doing so, DoD should

Figure 4.2
Process for Analytic Support for Strategic Planning

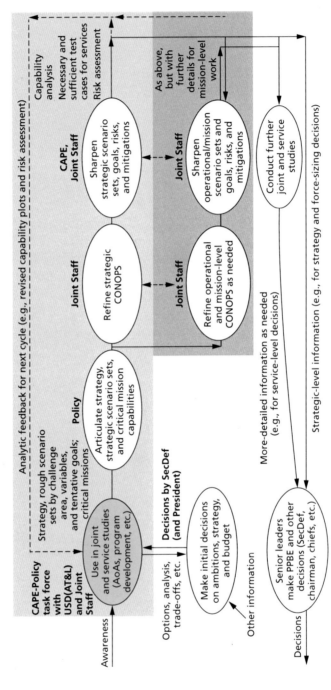

NOTES: Light and dark gray areas indicate the domain of ASSP; work in the darker area is somewhat more detailed. Policy = Office of the Under Secretary of Defense for Policy; PPBE = planning, programming, budgeting, and executing; SecDef = Secretary of Defense.

RAND RR1469-4.2

elevate the role of the Office of the Under Secretary of Defense for Acquisition, Technology and Logistics (USD[AT&L]) and have the Joint Staff be more active in ensuring the surfacing of innovative service options (not just programs of record and traditional CONOPS).[5]

4. ASSP should help OSD focus on strategic considerations[6] and relatively low-resolution analysis when establishing planning scenarios or studying mission-level issues (also called *capability issues*), with the Joint Staff having the primary role for the next level of detail that is especially important to the services.

5. DoD should require the Joint Staff to be more active in reviewing and critiquing service programs and requirement estimate.

6. ASSP should change the mix of analytic methods, increasing the emphasis on lower-resolution analysis with relatively simple qualitative models, quantitative models, *and* human wargaming. These should be sound abstractions from more-detailed (higher-resolution) work. ASSP should apply this mixed-method approach to both campaign- and mission-level analyses.

Early Decisions with Spanning Sets as Outputs

The first step in Figure 4.2 is to inform initial decisions about strategy and budget. The primary issue is what challenges the United States will take on in developing its defense program. The more challenges the United States takes on, and the higher the goals set for each, the higher the budget that is necessary. Achieving some goals is implausible even with very high budgets. Informing such decisions requires analysis that is understandable, broad in scope, and helpful for making strategic choices despite deep uncertainty. Such choices balance across objectives and risks.

If the decisions determine requirements for the military services, what requirements are needed, and how should they be expressed? Using concrete scenarios is traditional for DoD. The concreteness encourages enthusiastic problem-solving, in-depth thinking, and communication. The issue is deciding on which and what kind of scenarios. In the mid-1990s, it was a major and welcome change when DoD began using a number of *type scenarios* (e.g., defeating a regional rogue, defeating a near-peer regional power, or simultaneous conflicts).[7] DoD has since added *combination challenges* (e.g., dealing with attacks on the homeland while being engaged abroad in two theaters).[8] Each such type scenario could arise in diverse ways, demanding different U.S. capabilities. A principle for evaluating a force-planning strategy is that analysis must cover the variations: Having the capability to do well across the range of reasonable possibilities matters, rather than being able to do very well for a single base case. The principle has been expressed by secretaries of defense for many years.[9]

As demonstrated in modern work,[10] coherently dealing with the variations can be accomplished with a well-chosen spanning set of scenarios; capability to deal with those scenarios implies capability to deal with in-between or lesser-included possibilities. This is not planning for the worst case because no such worst case exists. A case that is the worst in some respects (e.g., short warning) will be easy in others (e.g., the adversary will also have fewer forces than after a mobilization). Consequently, if requirements are to be expressed as test-case scenarios, many such cases (a spanning set) are needed for each type scenario. Defining these is straightforward for analysts but not so for committees. Systematizing such work will be a natural but important extension of the analysis for the 2010 QDR, which developed different spanning sets to test overall force structure by stressing the forces in different ways.

Figure 4.3 contrasts old and new approaches. Under SSA, an analytic baseline was a single detailed example (the ovals) of a type scenario or challenge. In the new approach, each such challenge type is represented by a *set* of low-resolution cases (the circles), presenting different stresses for U.S. capabilities.[11] Because the cases are low resolution in character, all of them can be examined analytically with suitable models, including parametric variations. As illustrated schematically on the right of Figure 4.3, results of such exploratory analysis can be presented in *capability maps* showing for which cases capabilities are *sufficient* (the green region means success expected), *risky* (the yellow region means perhaps adequate for deterrence at best), or *inadequate* (red region means corresponding to probable failure). Good analysts can develop these maps based on broad considerations and low-resolution models; analysts can then identify test cases (indicated

Figure 4.3
Single Type Scenarios, Spanning Sets, and Capability Maps

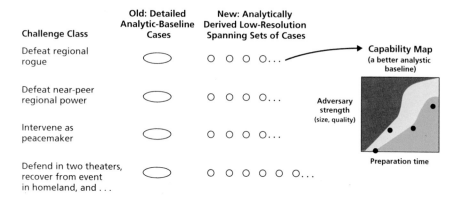

NOTES: Green, yellow, and red correspond, respectively, to probable success, uncertain results, or probable failure. Bullets indicate test cases, which constitute the spanning set.
RAND RR1469-4.3

by bullets) that stress the future force in the different ways needed and that constitute the spanning set. Those sets can be studied in more detail, improving and enriching the capability maps. These selective, detailed looks are sometimes crucial.

Why is this important? Why should DoD not just have one illustrative scenario of each class and assume that the services will build capabilities for all the variants? Arguably, during the Cold War, U.S. defense budgets were large, prices were lower, and the services did include just-in-case capabilities and slack. That is no longer the case given the budgetary strains on the U.S. military currently noted with alarm in the 2014 QDR and its critique (Hagel, 2014; Perry and Abizaid, 2014). Nonrequired capabilities and slack are seen as luxuries as the services scrimp in developing their programs while also attempting to protect such organizationally favored features as end strength, number of traditional units, and buy levels for programs of record. Thus, DoD needs to be more explicit about the breadth of requirements, or some needs will not be met.

Although the colored capability map in Figure 4.3 is simplified and has only two dimensions of uncertainty (preparation time and adversary strength), the basis for more-complex uncertainty analysis has been laid over with concrete examples. The approach is not hypothetical and can be comprehensibly extended to consider many more low-resolution dimensions of uncertainty (e.g., aggregate-level characterizations of the behavior of U.S. allies and adversaries, the operational strategies adopted, the real-world effectiveness of new weapon systems, and the effect of such shocks as initial cyber attacks on command and control). Although now well demonstrated, such analysis requires different methods and models than those used in the SSA activity, as well as people skilled in using the new methods and models.

Balancing Mission- and Campaign-Level Analyses

Another objective for ASSP is elevating the relative emphasis on mission-level work (also called *capability-area work*). Many crucial defense-planning issues are about how to achieve capabilities for accomplishing such increasingly difficult missions as defeating integrated air defenses. This is evident from even a cursory skim of recent QDRs. The ASSP activity should achieve a balance that gives at least as much, if not more, attention to mission-level issues. OUSDP and CAPE should primarily be concerned with higher-level (lower-resolution) aspects of both, while the services must address such matters in more depth. Precisely the same methods as described above apply well to mission-level analysis.

Despite the challenges in doing so, ASSP should also achieve a balance with respect to types of conflict. In particular, ASSP must give appropriate weight to analysis relating to counterterrorism, stabilization, and irregular warfare. The methods for studying these matters should draw on social science (and, in some cases, be the methods of social science).

Implications for Organization

Participation

ASSP should have the same membership as the previous analytic agenda and SSA activities (see Appendix B): OUSDP, CAPE, USD(AT&L), the Joint Staff, the services, and others as needed. However, some activities, particularly those informing interim decisions regarding strategy and defense scenarios while accounting for resource constraints, will continue to be more tightly limited to OSD and the Joint Staff. Whether those activities should be regarded as part of ASSP is a separate decision.

Responsibilities

Functional Issues

As summarized in Table 4.1, the new construct indicated in Figure 4.2 changes the responsibilities of DoD components. The first step of developing strategy and defense-planning scenarios is replaced by the larger step of first informing interim decisions on strategy and budget. This should be accomplished by an elite team or task force (not a committee) led by CAPE and OUSDP—CAPE because the effort requires rigorous, low-resolution analysis involving capabilities, costs, and trade-offs and OUSDP because the effort is about establishing strategy.

Table 4.1
Responsibilities in the Analysis for Support of Strategic Planning Activity

	Inform Initial Decisions on Strategy and Budget (Task Force Co-Led by Cape and OUSDP)	Articulate Strategy, Challenges, and Spanning Sets for Challenge	Refine Strategic CONOPS	Refine Strategic Capability Requirements and Assessments of Capability and Risk	Refine Detailed Requirements and Assessments of Capability and Risk as Needed
OUSDP	●●●●	●●●●	●	●	●
Joint Staff	●●	●	●●●●	●●	●●●●
CAPE	●●●●	●	●●	●●●●	●●
USD(AT&L)	●●	●	●●	●●	●●
SecDef	Approval	Approval		Approval	
VCJS, J5, or J8			Approval		Approval

NOTES: The task force activity (left side) might or might not be considered part of ASSP per se. Number of bullets indicates relative responsibility. SecDef = Secretary of Defense; VCJS = Vice Chairman, Joint Staff; J5 = Director for Strategic Plans and Policy, Joint Staff; J8 = Director, Force Structure, Resources, and Assessment, Joint Staff.

DoD should consider having some top-quality analysts functionally dual-hatted for CAPE and OUSDP (see below). Representatives from USD(AT&L) and the Joint Staff should ensure attention to innovation and military feasibility. Results of this crucial early effort should be approved by the Secretary of Defense and (ultimately) the President. To be sure, such an organizational change might be difficult, but the proposal has precedent.

The next step shown in the table is that OUSDP and the Joint Staff articulate strategy and develop CONOPS. The last two columns are new, corresponding to the lightly and more darkly shaded regions of Figure 4.1. The more-strategic, low-resolution activities are primarily the responsibility of CAPE, with the Secretary of Defense approving the final expression of requirements; those of a more detailed nature are primarily the responsibility of the Joint Staff. Although all ASSP contributors would review and comment on everything, as occurs within SSA, Table 4.1 indicates with two bullets that it is especially important for the Joint Staff (both J-5 and J-8) and USD(AT&L) to be involved in the low-resolution refining and that CAPE and USD(AT&L) should be involved in the more detailed work. The goal should not be consensus but ensuring quality and innovation.

One aspect of the new construct worth noting is that the later portions of the ASSP process (the part most akin to the current SSA) would have new responsibilities for refining capability maps and the related matter of risk assessment. In a sense, SSA's detailed analytic baselines for point cases would be replaced by the ASSP's lower-resolution capability maps.

Dealing with Organizational Tensions

How might the CAPE-OUSDP task force shown in Figure 4.2 be achieved? Possibilities include having a group within OUSDP effectively dual-hatted, responsive to and working closely with both the Under Secretary of Defense for Policy and the director of CAPE; having a group within CAPE effectively dual-hatted in the same way; or having the task force report directly to the deputy secretary.

Locus in OUSDP

It might be argued that the basis for the first organization already exists in OUSDP as the Assistant Secretary for Strategy, Plans and Capabilities (particularly the Strategy office beneath it). That office (see Figure 4.4), however, does not currently see itself as dual-hatted or as serving functions in both the OUSDP and CAPE lines.

Locus in CAPE

In earlier decades, the predecessor of CAPE, Program Analysis and Evaluation (PA&E), had a deputate called Regional Programs, which routinely contributed to the functions of OUSDP and PA&E. Regional Programs generated the defense-planning scenarios and influenced programs. The deputate's analysts were seen as both program analysts and strategic analysts. In the 1990s, DoD reorganized and created the ASD for Strategy and Resources, the predecessor of today's Assistant Secretary of Defense (ASD) for Strategy, Plans and Capabilities. The office took on responsibilities for building strategy and related scenarios. In my observation, the organizational transition, which seemed logical at the time, was not fully successful because the office had never been very influential in program development.[12] Nor did it have a sufficient cadre of analysts respected across DoD for program analysis to complement its expertise in strategy analysis.

Figure 4.4
Strategy Office Within OUSDP

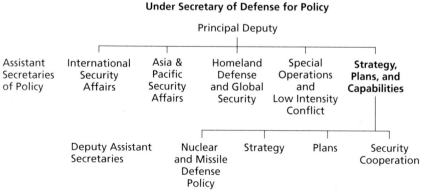

Locus in OSD

It might be that the dual-hatting would not work or that CAPE would be unwilling to reassume the responsibility for leading DoD-wide analysis. CAPE perhaps has this responsibility under its charter, but that responsibility is the third priority, well behind program evaluation and cost assessment. During my research for this study, observers I spoke with noted that CAPE no longer supports or participates in the professional society promoting military operations research and seems to have little interest in leading DoD-wide analysis. That could change, especially if necessary billets were added, but the last option would be to have the functionally dual-hatted organization report to the deputy secretary. That would be natural in some respects and might work well with some combinations of personalities, but it would expand the office's already large span of control and probably cause the organization to have a tense relationship with both OUSDP and CAPE.[13]

A Portfolio of Supporting Activities

The previous section discussed organizational structure at an aggregate level: what offices must be involved, who should have what responsibilities, and how information-sharing should be achieved. A separate issue is the portfolio of activities. Historically, the analytic agenda and SSA were dominated by campaign-modeling, with some important exceptions.

In light of the discussion in Chapter Three on methods and long-standing shortcomings, the recommendation here is that DoD should conceive of its ASSP analytic activities in portfolio terms, as suggested in Figure 4.5.

This approach would push back against the common tendency of analytical organizations to focus on increasingly detailed computer models over time with a continuing and expensive search for agreed-on data and refinements, while putting little investment into research, multiresolution modeling, or integration of knowledge across sources and levels.

On the one hand, an organized approach such as that suggested for ASSP would fill in gaps and provide integrated analytic capabilities. It is also possible, however, that it would crowd out ad hoc proj-

Figure 4.5
An Activity Domain for Supporting Activities

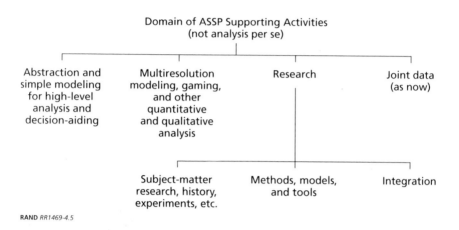

RAND *RR1469-4.5*

ects by various DoD components that might be of higher quality and more directly relevant than those produced by a centralized activity. From time to time, DoD officials have attempted to centralize and systematize modeling; the results have typically been bad. Thus, any structure such as that displayed in Figure 4.5 should insist on diversity and competition, rather than pursuit of some allegedly cost-effective standard. Research, modeling, and analysis are not like commodities except perhaps when the phenomena they deal with are well understood and stable.

Implications for Methods and Tools

Legacy Methods and the Broader Landscape

Campaign-modeling dominated SSA until 2011. Some human wargaming has been added in recent years. The spectrum of relevant methods is actually much broader. Table 4.2 is one version of a table that appeared in reports and briefings over time.[14] The instruments range from simple back-of-the-envelope models to field experiments. Although incomplete, the list broadens the scope of thinking about what tools the analyst can use. The columns describe various attri-

Table 4.2
Illustrating the Broad Range of Analytic Instruments

Instrument	Resolution	Relative Strength — Strategic-Level Functionality						
		Agility, Creativity, Transparency	Breadth	Strategic Decision-Aiding	Strategic Integration	Physical Phenomena	Human Phenomena	Empirical Cautions
Simple analytical[a]	Low	5	1	5	1	1	1	N.A.
Seminar-level human wargaming	Low	5	4	3	1	1	4	3
Red-teaming on capabilities and operations	Varied	5	3	3	1	3	5	5
Qualitative factor trees	Low	5	5	5	5	1	3	3
Human wargaming	Medium	1	5	5	5	3	3	5
Campaign simulation (usual)	Medium	2	5	2	4	2	1	N.A.
+ agents, political-economic factors, and exploratory analysis[b]	Medium	3	5	4	5	2	3	N.A.
Mission-level adaptive models, exploratory analysis	Medium	3	3	3	1	5	3	N.A.
High-fidelity simulation[c]	High	1	1	1	1	5	1	3
Historical case studies	Varied	1	1	1	1	3	5	5
Historical data analysis	Low	1	1	1	1	1	1	5
Field experiments and war data	Varied	1	1	1	1	5	5	5

SOURCE: Davis, 2014, Table 3.1.

NOTES: Ratings are 1 (very poor) to 5 (very good), with red, orange, yellow, light green, and green corresponding to 1, 2, 3, 4, and 5, respectively. Scores depend on assumptions.

[a] Examples include closed-form models and spreadsheet-level computer models.

[b] Exploratory analysis examines the effect of simultaneous variations of all important assumptions, not mere sensitivity analysis on the margin.

[c] In some instances, high-fidelity simulation can be a primary and reliable source of what can be considered to be empirical information. It is simply not feasible to obtain the equivalent information with physical testing.

butes of the methods. Although the scoring is subjective and dependent on detailed assumptions (see Davis, 2014), the primary point is that some methods tend to be associated with low resolution, agility, creativity, and transparency (top left), while others are associated with reality (bottom right). No single method is good across the board, and, to make things worse, typical campaign simulations have been much weaker in important attributes than they need to be.

DoD should arguably be using *all* of these methods and sources of information, as well as some not on the list, as contributors to ASSP. We ignore history at our peril, as was demonstrated in 2003, when U.S. forces employed in Iraq proved inadequate for the insurgency that arose. History-based analysis had pointed out the potential need for much larger forces, but this idea had been disparaged or sloughed off. As a second example, the asymmetric tactics being used by the Islamic State of Iraq and the Levant (ISIL) and other adversaries today have come as no surprise to history-minded analysts but were not anticipated in campaign studies of the traditional variety. A contrasting example: In the 1990s, I remember well campaign analyses that used history-based parameter values to determine the effectiveness of air forces grossly underestimated the effectiveness of air forces in the age of precision fires, whereas other simple models coupled with test data proved prescient.[15]

At the higher levels of political-military analysis, seminar wargames and strategic red-teaming have a long history of anticipating developments (and, to be sure, sometimes those developments did not materialize) (Defense Science Board, 2003b). This is among the many reasons for the resurgence of interest in wargaming.

An entirely different class of analysis has arisen in response to the September 11, 2001, attacks by al-Qaeda and the subsequent global war on terrorism. Some of the tools have been variants of campaign models that include certain political-military aspects related to, for example, insurgency and its roots in societal discontent (Body and Marston, 2011). Others have been mainstream social science tools, such as methods for eliciting expert opinion, quantitative correlational work, forecasting, and case studies (see, e.g., Wong, 2014). Still others have contributed to an evolving approach to causal social science modeling about, for example, the factors contributing to terrorism and public support thereof. That work was sponsored by the DoD analytic community (many of the same people associated with SSA) and drew heavily on the academic social science literature and major contributors to that literature to integrate many of the seemingly inconsistent threads of research on counterterrorism.[16] In more-recent times, this work has been extended to generate prototypes of computational models that are based on qualitative social science research but designed to highlight uncertainties and facilitate analysis in inherently "soft" circumstances.[17]

Uncertainty Analysis

A theme throughout this report is DoD's need to do far better in addressing uncertainty, and to do so routinely in analysis rather than by exception. A great deal of modern research describes methods for doing so, with numerous published examples in a variety of social-policy and national-security domains. Such uncertainty analysis was not feasible decades ago when the legacy analytic methods were developed, but much progress has occurred. A few key points are the following:

- The methods apply not just to "normal" uncertainty (as when representing empirical data on reliability statistics) but to *deep uncertainty*, which has been referred to over the years as, for example, *real uncertainty*, *scenario uncertainty*, and possible *wild cards*. More technically, *deep uncertainty* has been defined as the condition in which "the parties to a decision do not know or agree on the system models relating actions to consequences or the prior probability distributions for the key input parameter to those models."[18] That situation is the norm in strategic planning.
- Addressing deep uncertainty with modern methods can aid in practical decisionmaking, rather than lead to the paralysis by analysis deplored by decisionmakers.[19] Such methods relate well to DoD's needs for capabilities analysis and strategic portfolio analysis. They work best, however, when used with relatively simple models or multiresolution families of models (Davis, Gompert, et al., 2008; Davis, 2014).

Human Wargaming

In May 2015, Deputy Secretary of Defense Robert Work announced a new initiative to use human wargaming as a major tool in contemplating needs for the years ahead,[20] distinguishing among roughly three time horizons: the first Future Years Defense Program (FYDP), the second and third FYDP, and the fourth FYDP and beyond. One part of this was to "reinvigorate Support for Strategic Analysis (SSA)" with the intention that OUSDP will conduct wargames to inform SSA scenario development, that the Joint Staff will conduct wargames to inform CONOPS and force development, and that CAPE will "manage the development of force capability and capability excursions in SSA scenarios to allow a broader exploration of risks and solutions."[21]

Since the announcement, a great deal of activity ensued as individuals and organizations reacquainted themselves with wargaming and contemplated how they could contribute. Further, DoD components conducted a large number of wargames. Authors have written thoughtful papers based on experience,[22] as well as earlier reviews of wargaming.[23] The results of one recent game illustrate timeliness and flexibility (Shlapak and Johnson, 2016). This game addressed the challenges faced by NATO in deterring or defending its Baltic-state members from possible aggression by Russia. It was an interesting example of how low-resolution work accomplished in a relatively short time can rather convincingly concentrate minds, rule out some options, and suggest a path ahead that is not sensitive to the kinds of details that must be dealt with in computer models. Such games often set the stage for subsequent work, as illustrated by several decades of experience with RAND's Day After wargames.[24]

An enduring challenge is how to use human gaming more rigorously, enough so that gaming can be seen as part of the analytic process. Views differ, but Figure 4.6 indicates schematically the approach highlighted in National Research Council, 2014.[25] This approach builds cumulative knowledge into models, which can then be used for transparent, reproducible, rigorous analysis. Gaming is a key mechanism for building the knowledge: identifying factors that human players immediately recognize but that initial modeling has omitted, gaining a better sense of how humans might act, and so on. This is a model-

Figure 4.6
Integrating Human Wargaming and Modeling

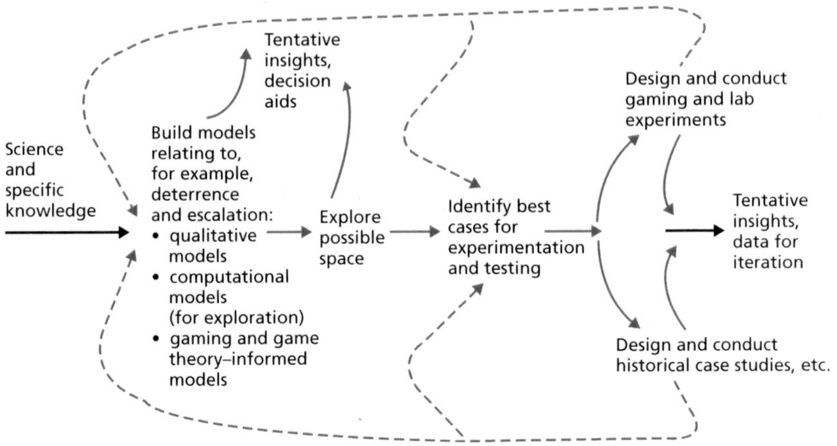

SOURCE: Adapted from National Research Council, 2014.
RAND RR1469-4.6

test-model approach.[26] The model informs game design; the game results inform model improvement. The example assumes an application studying issues of deterrence and escalation, such as in a possible conflict with North Korea. The model referred to is a simple *cognitive model* of North Korea, one that attempts to represent the factors that North Korean leaders would consider when contemplating escalation or deescalation (see National Research Council, 2014).

A very different approach regards gaming as involving low-resolution models adequate for certain important purposes, especially when results are overdetermined. The recent gaming on deterring Russian aggression in the Baltics is an example (Shlapak and Johnson, 2016).

Another unresolved problem is what should constitute *validation* for human gaming. A great deal of effort has gone into studying and writing about the validation of models, but much less has been done on this for human gaming. The primary need is to have first-class participants and enough funding to do things properly. Preparing a serious game can and should take significant time, whereas merely throwing together a quick game is deceptively simple. This is something with which several military institutions are well aware from decades of experience (e.g., the U.S. Naval War College). The Joint Staff has also conducted numerous high-level wargames for many years, many of them seminar-style. In recent years, FFRDCs have also conducted wargames for several of the DoD components on a variety of current issues.

Overall, the rediscovery of human wargaming is very much to be applauded. How well it fits into DoD's joint analysis activities is yet to be determined. DoD is proposing significant efforts of this kind in its budgeting, in part to revitalize the SSA activity.

New Campaign and Mission Models

Given controversy about campaign models, it might seem odd to suggest new developments, but campaign analysis is crucial for integration and for certain activities, such as testing the executability and appropriateness of a tentative strategy and associated programs, or identifying such problems as the double-counting of units and subtle interdependencies. These activities are useful in force-*sizing* analysis. The campaign analyses conducted within DoD over the past two to three decades have sometimes been associated with complexity, opaqueness, inflexibility, and failure to address uncertainty, but *none of these shortcomings is inherent.* Mission models can be even more detailed and (except to those who use them) opaque. They also, however, can be relatively simple, parametric, and suitable for higher-level analysis. They are useful in illuminating what some refer to as *capability* issues.

Since relatively detailed campaign and mission models already exist, a priority should be for DoD to develop simpler models and connections between those and the more-detailed models. Table 4.3 sketches alternative ways to develop these simpler models. All have precedents (Davis, 2014, Chapter Four). The scoring of the options depends on a number of questionable assumptions, so the reader should take away only the fact that options exist. An example of an empirical model would be CAPE's statistical models of system cost or potential cost growth. Qualitative models may use graphics (e.g., factor trees) or logic tables showing the factors that drive results. Simple models, started from scratch, are often spreadsheet-level in complexity. CAPE

Table 4.3
Alternative Paths to Simple Models

Path	Understandability[a]	Parametric Flexibility	Organizational Acceptance[b]	Development Risk[c]	Level of Effort[d]
Empirical model (e.g., regression)	•		••	••	••
Qualitative logic table by case (drawing on wargame experience)	••••	••	varied		••
Fresh simple model	••••	••••	varied	•	•
Simplify parts of existing models	••	••••	••••	•••	••
Develop motivated metamodels	••	••	••••	••	••
Build new multiresolution models or families	••••	••••	••••	••••	••••

NOTE: Number of bullets denotes, roughly, degree to which option achieves attribute of column.

[a] Regressions may be simple but not substantively revealing.

[b] Policymakers often accept clear analysis by trusted agents while organizations relying on big models resist.

[c] Development risk is highly variable and depends on quality of team, clarity of purpose, absence of consensus culture, and so on.

[d] Highly dependent on the quality of design and, when relating to existing models, the nature of the model.

is sponsoring a recent study that the author regards as building a low-resolution "campaign model" designed explicitly for far-ranging uncertainty analysis at the strategic level.[27] Simplifying parts of an existing model might mean, for example, turning a module of a large model such as STORM into a stand-alone model and eliminating some of its detail. A motivated metamodel is a statistical model fitted to a form suggested by an understanding of the phenomenon being viewed, rather than based on linear regression of input variables. Building new multiresolution models or families is straightforward for relatively simple problems (e.g., modeling close-air support's effectiveness as a function of a half-dozen variables) but more complex for something such as campaign models. Such models allow the user to decide on the level of detail to be entered as inputs or the level of detail to reason at.

Deciding which combination of these approaches should be taken goes far beyond the scope of this report and depends on developer teams, the detailed characteristics of the current models, and DoD's ability to manage work. Past experience on the last item is not encouraging. I note, as corroborated in my interviewing for this report, that the last major DoD development, the Joint Warfare System (JWARS), later called the Joint Analysis System (JAS), is largely considered to have been a fiasco. The system is estimated to have involved 500 person-years of effort over ten years and was almost never used for its intended DoD purposes (see Allen et al., 2007, p. 9). The JWARS development had numerous problems from the outset. In many respects, the system was designed by a committee, with the requirements stemming from lengthy wish lists from all stakeholders. A second problem was that JWARS was postulated to serve many functions with very different needs. Yet a third problem was that, although aspects of the software development were modern, the result lacked adequate modularity and multiresolution features.

Other model developments, such as those for IDAGAM, TACWAR, THUNDER, JICM, and STORM (the models are known by their acronyms), were developed by single organizations. The level of effort for the first four was at least 20 person-years but not 500 (Allen et al., 2007).

Campaign analysis for particular studies has often been accomplished with smaller, simpler, and highly parametric models focused on the particular issues of interest (e.g., strategic mobility, attacking any invading mechanized armored forces with precision weapons, ballistic-missile defense, the air campaign against a peer competitor's fixed targets). Such analysis is sometimes described as mission-level analysis, but it is actually module-focused campaign analysis, still at relatively low resolution and highly parametric. To give one simple example, for long-range precision fires to halt an invading mechanized force moving along major roads, results can be well understood with simple models involving, for example, sorties per day, kills per sortie, movement speed of the army, the army's break point (the level of attrition beyond which movement can be assumed to stop), and a few other parameters (Ochmanek et al., 1998; Davis, McEver, and Wilson, 2002).

It is of interest that a community review in 2007 (Davis and Henninger, 2007) recommended a new approach to theater modeling that would emphasize modularity, competition among module developers, uncertainty analysis, and so on.

Analytic Tools for Counterterrorism, Stabilization, and Irregular Warfare

Among the more-challenging issues for DoD joint analysis are those posed by counterterrorism and irregular warfare. It is unclear to what extent such issues are amenable to modeling, but some advances have been made in qualitative modeling (Davis and Cragin, 2009) and even in extending social science–based modeling to a special kind of computational modeling that emphasizes uncertainty and frames for thinking rather than prediction (Davis and O'Mahony, 2013). Human gaming has proven useful.

Implications for Staffing

It is unlikely that DoD will want to do a whole-scale review of its staffing for joint analysis. Many excellent practitioners are in place. However, changes are possible and desirable. DoD should plan simultaneously for a mix of in-house and FFRDC capabilities (with additional but crucial capabilities in laboratories and industry). Appendix D discusses briefly some options for DoD to consider on the specific matter of where to develop, maintain, and do joint campaign and mission-level modeling.

Misconceptions
Who Are Analysts?

A parochial interpretation of the term *analyst* is that analysts do whatever it is that CAPE and corresponding component offices do, which is sometimes referred to as number-crunching, running big models, and budgeting. In fact, analysts are distributed throughout DoD and related contract organizations. Analysts perform many very different kinds of functions. For example, USD(AT&L) and OSD for Personnel and Readiness need a great deal of analysis, as does OUSDP, although its analysis is less quantitative in nature. Within the Joint Staff, analysis occurs in many of the offices.

Even with CAPE, most analysts are not the analysts associated with the SSA activity. Instead, they work in particular divisions dealing with, for example, ground, air, or land forces; strategic forces; or cyber war—usually at the level of specific programs. Somewhat the same is true, but with important differences, for analysts in Air Force Studies and Analysis, the Navy's N-81, the Marine Corps Combat Development Command, and the Army's G-8 and Center for Army Analysis offices.[28] Many of the analysts outside the SSA activity at least think and work in the joint context, if not always with an appreciation of subtleties.

What Are the Backgrounds of Good Analysts?

Even when referring to the DoD joint-analysis domain, the term *analyst* should not be interpreted as synonymous with *someone with an operations research degree*. It is merely an artifact of bureaucratic processes over time that this misconception has arisen (services assign specialty codes to their personnel, based in part on degrees achieved, and operations research is one of them). Taking a broader view, top-notch analysts—even those capable of quantitative or otherwise rigorous analysis—have been educated in a broad range of disciplines, including the physical sciences, economics, engineering, mathematics, applied mathematics (e.g., operations research), computer science, law, and political science. Indeed, those often identified as giants in the pantheon of past DoD analysts have only sometimes been trained in operations research. Appendix C discusses backgrounds of notable analysts and of those hired in some analysis institutions. The primary conclusion is one of diversity in disciplinary background.

CHAPTER FIVE

Recommendations

Much of Chapter Four amounts to recommendations, but it is well to synopsize the recommendations here. This report recommends that DoD do the following:

- Adopt a new planning construct (see Figure 4.2) with new organizational responsibilities (see Table 4.1).
- Revamp the analytic methods used, taking a family-of-methods approach that includes a mix of lower- and higher-resolution modeling and human gaming and integration thereof. This approach should increase the emphasis on lower-resolution methods.
- Rebalance the mix of staff and use of partner organizations, such as FFRDCs, accordingly.
- Continuously invest in research to support the above analyses with a broad range of information, both qualitative and quantitative, and develop and disseminate new or refreshed methods and tools, also both qualitative and quantitative. The quality of joint analysis depends on the methods evolving and adapting over time.

These recommendations come with an important caveat. Halfway measures in implementation might be costly and burdensome while accomplishing little. In particular, it will be essential to change cultural patterns that have hurt DoD's joint analysis. These patterns include a counterproductive focus on standard cases rather than planning for broader capabilities; a tendency to uncritically extrapolate current force structures in PPBE-related analysis; a failure to exploit new technology and concepts of operation and organization; and excessive dependence on large, complex models ill-suited for planning under uncertainty.

Congressional Request

Section 1053 of the 2015 defense authorization bill called for a study with the following elements (U.S. Congress, 2014):

(a) INDEPENDENT ASSESSMENT. The Secretary of Defense shall commission an appropriate entity outside the Department Defense to conduct an independent assessment of the joint analytic capabilities of the Department of Defense to support strategy, plans, and force development and their link to resource decisions.

(b) ELEMENTS. The assessment required by subsection (a) shall include each of the following:

(1) An assessment of the analytical capability of the Office of the Secretary of Defense and the Joint Staff to support force planning, defense strategy development, program and budget decisions, and the review of war plans.[1]

(2) Recommendations on improvements to such capability as required, including changes to processes or organizations that may be necessary.

(c) REPORT. Not later than one year after the date of the enactment of this Act, the entity that conducts the assessment required by subsection (a) shall provide to the Secretary an unclassified report, with a classified annex (if appropriate), containing its findings as a result of the assessment. Not later than 90 days after the date of the receipt of the report, the Secretary shall transmit the report to the congressional defense committees, together with such comments on the report as the Secretary considers appropriate.

Support for Strategic Analysis

The 2001 QDR, issued shortly after the September 11 attack by al-Qaeda on New York and the Pentagon, has strong language dictating changes in defense planning:

> The approach shifts the focus of U.S. force planning from optimizing for conflicts in two particular regions—Northeast and Southwest Asia—to building a *portfolio of capabilities that is robust across the spectrum of possible force requirements*, both functional and geographical. This approach to planning responds to the capabilities-based strategy outlined above. It focused more on how an adversary might fight than on who the adversary might be and where a war might occur. The shift is intended to refocus planners on the growing range of capabilities that adversaries might possess or could develop. It will require planners to define the military objectives associated with defeating aggression or coercion in a variety of potential scenarios in addition to conventional cross-border invasions. *It calls for identifying, developing, and fielding capabilities that, for a given level of forces, would accomplish each mission at an acceptable level of risk* as established by the National Command Authorities. (Rumsfeld, 2001, p. 17; emphasis added)

The Analytic Agenda

In one of the follow-up actions to the QDR and the related Defense Planning Guidance, DoD issued a directive creating an activity, titled *Data Collection, Development, and Management in Support of Strategic Analysis* (DoD Directive 8260.1, 2002). This came to be called the *analytic agenda*. The purpose was to establish policy and assign responsibility for "generating, collecting, developing, maintaining, and disseminating data on current and future U.S. and non-U.S. forces in support of strategic analysis conducted by the Department of Defense."

Anticipating that a number of planning scenarios would be used, the directive called for the construction of analytical baselines for each scenario. Such an analytic baseline would include a scenario, CONOPS, and integrated data to be used as a "foundation for strategic analyses" in computer-assisted wargames and theater campaign simulations (DoD Directive 8260.1, 2002).

A *scenario*, in this context, meant a synopsis of the road to war and a projected course of action at the strategic and operational levels of warfare. The synopsis would include political-military contexts, assumptions, operational objectives, major force arrivals, and planning considerations. Strategic analysis would then be accomplished to assess the ability of current or planned forces to address those scenarios given priority—i.e., to "execute the defense strategy" (DoD Directive 8260.1, 2002).

The primary responsibilities were identified as follows:

- OUSDP: develop and establish priorities among scenarios set in future periods and build scenario descriptions, including road-to-war and planning factors (e.g., warning time, concurrency, assumed postures of engagement)
- Chairman of the Joint Chiefs of Staff (in coordination with services): develop baselines for use of *current* forces
- Director of PA&E (in coordination with the services): develop baselines of *future* forces, threats, and scenarios; establish a management structure and data repository

- service heads: support OSD and Joint Staff activities with data and advice
- combatant commanders: provide operational advice when the scenario was in their areas of responsibilities and generate current-year analytical baselines using operational plans
- intelligence community: provide projections of adversary future force capabilities and capacities.

Regrettably, in light of the intentions of the secretary, the directive said nothing explicit about uncertainty. Although the term *baseline* meant the base from which analysis would proceed, the imperative to consider extensive excursions was not mentioned explicitly.[1]

The subsequent directive (DoD Directive 8260.2, 2003) was about details of implementation but also included some clarifications and modifications. For example, the directive specified that the Chairman of the Joint Chiefs of Staff would, in coordination, with the services and with the director of PA&E, "prepare annually an integrated multiyear program for developing analytic baselines for use in strategic analyses, based upon scenario priorities identified by the [Under Secretary of Defense for Policy]." Thus, the role of the Chairman of the Joint Chiefs of Staff would not be limited merely to considering CONOPS for current forces.

Again, the directive said nothing to encourage uncertainty analysis except in referring to analytical baselines as "starting points" for analysis, which might have been read to imply extensive excursions but was largely read to mean *starting points and end points*.

Early Description of Analytic Agenda Intentions

A rather detailed account of intentions for the analytic agenda, which became the SSA activity after 2010, was provided in 2004 by CAPE's Deputy for Force Structure and Risk Analysis:

> A new DoD initiative—capabilities based planning—shifts planning focus from the performance of systems across a relatively narrow range of threats to the achievement of specific objectives in response to *a broad set of robust and adaptive threats*. It applies to both current and future years, and makes trade-offs across

alternative capabilities to accomplish objectives. Capabilities-based planning will help frame the debate for allocating scarce resources among diverse sets of alternatives. It encompasses most DoD planning activities, including:

. . . Future Force Planning. Helps senior decision makers develop defense programs that provide *robust capabilities, while minimizing risk* within a constrained resource environment. . . .

[S]everal activities . . . will improve our ability to perform analysis and hence support capabilities-based planning. One is the year-old DoD-wide initiative called the "analytic agenda." Its purpose is to improve the quality and timeliness of analysis, with a scope ranging from more comprehensive scenarios to better data and analytical tools. The end result will be more consistent and visible data and analyses. The components of the analytic agenda—including Defense Planning Scenarios (DPSs), Multi-Service Force Deployment (MSFD) data sets, and Analytical Baselines (ABs) for analyses using future forces, along with operations plans and ABs for analyses using current-year forces—are discussed below.[2]

The underpinning of the analytic agenda is the establishment of a series of DPSs. These scenarios represent a set of realistic mid-term and far-term challenges that could require a US military response. DPSs differ from the Illustrative Planning Scenarios of the past in that they explicitly acknowledge uncertainties surrounding our ability to precisely describe our future adversaries.

The MSFD provides, for each DPS, order-of-battle detail for US, allied, and opposing forces. It includes the initial location of units, the types of equipment they possess, and how they plan to operate. MSFD development is led by the Joint Staff, with the involvement of the intelligence community, the military services, and the combatant commanders.

A completed MSFD establishes the initial conditions for a postulated conflict. An example would be the initial unit locations and plans for a future Iraqi Freedom-like operation; another would be similar information for a smaller-scale contingency, such as our operations in Bosnia. We then run this data through our analytic tools to forecast the results of the conflict. We carefully review all aspects of the conflict—including deployments, logistics and engagements—to ensure that the data and models are producing defensible results. After we are satisfied with the analysis, we package the associated data (DPS, MSFD, and model with inputs and outputs) into a product called an Analytical Baseline (AB). ABs provide consistent and visible starting points for future analyses. They also greatly increase our ability to generate timely answers.

A similar process, using combatant commanders' operational plans in place of the DPS and MSFD, develops ABs for current-year scenarios. Over the next several years, this process will establish five to seven current-year ABs, and an equal number of future-year ABs as starting points for analysis. In addition, for some scenarios (e.g., some small scale contingencies and homeland defense military actions) only the DPS and MSFD will be needed. No matter what scenario or type of operation is being explored, a valuable by-product of the analytic agenda process is a common language for discussing and comparing analyses. At full maturity, *it is anticipated that most analyses reaching senior leaders will have been shared and vetted through the analytic agenda process. Ultimately, we expect these common baselines to permit analytical excursions that will unveil a wide range of new opportunities to defense planners.*

To implement this process, the Department issued guidance (DoD Directive 8260.1 and DoD Instruction 8260.2) establishing a common data-sharing environment. The Joint Data Support (JDS) office has been chartered as the single repository for all analytical agenda data. (Coulter, 2004; emphasis added)

An Early Failed Attempt at a Strong Interpretation

As discussed during my interviewing for this project, in 2002–2003, OUSDP attempted to mandate an interpretation in which SSA work would systematically explore capabilities for numerous versions of each type of scenario (e.g., many versions of a possible conflict with a regional power or of a conflict with a near-peer power). The desire was to generate information about the significance of such crucial assumptions as strategic warning, separation in time between conflicts, and the adversary's objectives and strategy. The initiative was abandoned because those responsible for joint analysis insisted that it was not feasible to do such exploration. The fundamental problem was that the Joint Staff's development of the Multi-Service Force Deployment required manpower-intensive efforts, by operations-savvy officers, that sought to develop realistic and responsive CONOPS. Doing so for each scenario variant was deemed implausible. Instead, the pattern became one of the Joint Staff building a well-conceived CONOPS for a single version of the scenario, albeit with a dutiful listing of the many corresponding assumptions. In effect, the so-called analyst community (essentially those associated most closely with SSA) claimed an inability to accomplish the kind of analysis sought by leadership. This community continued to do analysis much as it had in the 1990s, using mostly the same models and methods. A major difference was that everyone was able to use the same joint data (not easily possible in the 1990s).

Viewgraph-level discussion obscured some of the problems. After all, DoD was now considering an increasingly large number of type scenarios and thereby taking a much broader view than it had until the mid-1990s. Indeed, it sometimes seemed that DoD was considering more scenarios than the system could deal with well. The conceptual mischief here was that *the* scenario used for a given type of conflict (e.g., major theater war with a regional power) was only one of an infinite variety. Although some might imagine that this would suffice if the scenario represented a worst plausible case, no such single scenario exists because a scenario posing a serious challenge in some respects (e.g., short warning) necessarily poses only a modest challenge in other respects (if the United States has short warning, then the adversary is unlikely to have had substantial time for mobilization). As another example, an adversary that is threatening and seriously intending to use nuclear weapons to reestablish deterrence if the United States intervenes poses different military problems than one pursuing a purely conventional strategy that might eventually escalate. As a third example, suppose that certain U.S. capabilities are dramatically greater than nominally assumed or, to the contrary, that certain adversary capabilities are dramatically better than in the current best estimate (e.g., vulnerabilities projected for an integrated air-defense system vanish). To understand the capabilities needed for any such single type scenario, it is necessary to consider a wide range of variants, something recognized since the 1980s (Davis, 1994, 2002). Doing so requires different models and analytic methods than those used in the activities of the analytic agenda and SSA (see the discussion in Chapter Four). While SSA's models could and were used for numerous excursions by CAPE's SAC, those were primarily excursions that were incremental in nature, rather than ones posing distinctly different challenges and, presumably, different operational strategies and related CONOPS.

SSA 2011

A new directive emerged in July 2011 (DoD Directive 8260.05, 2011). This specified that

- SSA products shall include (1) current baselines reflecting combatant commander plans and approved force-management decisions, and (2) near- to long-term scenarios, CONOPS, forces, and baselines based on plausible challenges requiring DoD resources and capabilities
- product development shall be a collaborative and iterative process led by CAPE, OUSDP, and the Joint Staff
- unresolved issues shall be referred to the Secretary of Defense.

An enclosure to the directive provided somewhat more detail. For example, scenario development is referred to as a responsibility of OUSDP, but in collaboration with the director of CAPE and the Chairman of the Joint Chiefs of Staff and in coordination with the heads of OSD and other DoD components. The Chairman of the Joint Chiefs of Staff will "manage the development of and approve SSA CONOPS and forces."

Figures B.1 and B.2 show SSA membership and the pre-2012 SSA process, respectively.

Figure B.1
Participants in Support for Strategic Analysis

DoD Analysis Community
(planning, programming, acquisition)
Guided by DODI 8260.01

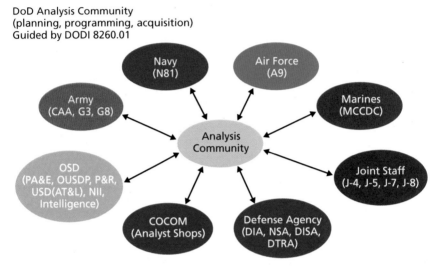

SOURCE: Stevens, 2008. Later versions of this briefing noted inclusion of the
interagency community, industry and FFRDCs, and allies.
NOTE: CAA = Center for Army Analysis; COCOM = combatant commander;
DIA = Defense Intelligence Agency; DISA = Defense Information Systems Agency;
DODI = Department of Defense instruction; DTRA = Defense Threat Reduction
Agency; MCCDC = Marine Corps Combat Development Command;
NSA = National Security Agency; NII = Networks and Information Integration;
P&R = Personnel and Readiness.
RAND RR1469-B.1

Figure B.2
A Depiction of the Support for Strategic Analysis Process (as of 2011)

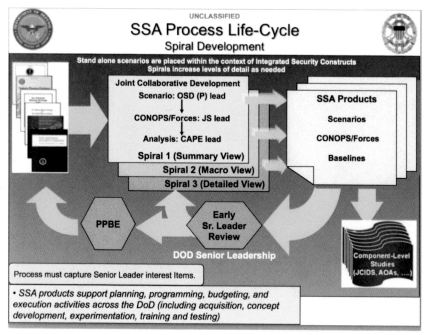

SOURCE: Excerpted from Keener, 2011.
RAND RR1469-B.2

The Varied Backgrounds of Senior Department of Defense Analysts

In the course of this research, I was asked about the disciplinary backgrounds of notable DoD analysts over the years. Tables C.1 and C.2 give two listings. Table C.1 shows the backgrounds of those who have led CAPE or its predecessor organizations. Table C.2 shows recipients of the most prestigious award of the Military Operations Research Society, the Vincent Wanner Award. The primary observation is that the range of backgrounds is substantial and, in particular, is not particularly associated with applied mathematics or operations research.

For some further data, Table C.3 shows the range of backgrounds of RAND staff with doctorates. Roughly half or more of RAND staff work on non-DoD problems, but the same range (if not the percentages) exists for national-security work (see Figure C.1). Figure C.2 is a roughly comparable chart for the Institute for Defense Analyses, which is much more heavily dedicated to national-security work.[1] As a final data point, the CAPE website states that

> CAPE's civilian and military staff is highly educated, with backgrounds in a variety of academic disciplines, including physics, economics, engineering, mathematics, biology, computer science, and operations research. Approximately 35% of the analytic civilian staff holds a doctorate in their field, and a substantial share have prior military experience. (CAPE, undated)

Table C.1
Disciplinary Backgrounds of CAPE Leadership

Name	Discipline	Date
Alain Enthoven	Economics	1965–1969
Ivan Selin	Engineering	1969–1970
Gardiner Tucker	Physics	1970–1973
Leonard Sullivan	Engineering	1973–1976
Edward C. Aldridge	Engineering, science	1976–1977
Russell Murray	Engineering	1977–1981
David C. Chu	Economics	1981–1993
William J. Lynn	Law, public affairs	1993–1997
Robert R. Soule	Economics	1998–2001
Barry D. Watts	History	2001–2002
Stephen A. Cambone	Political science	2002–2003
Kenneth J. Krieg	Public policy	2003–2005
Bradley M. Berkson	Engineering, business administration	2005–2009
Christine H. Fox	Mathematics	2009–2014
Jamie M. Morin	Public policy and administration, political science, engineering	2014–present

NOTE: At different points, the office has been called Systems Analysis, Program Analysis and Evaluation, and Cost Assessment and Program Evaluation.

Other contributors to SSA (such as those in OUSDP) typically have backgrounds in political science, international relations, and other social science subjects.

Table C.2
Winners of the Vincent Wanner Award

Name	Disciplinary Background	Date
Mr. Ervin Kapos	Mathematics	2015
Mr. Jack F. Keane	Operations research	2014
Dr. Gregory S. Parnell	Engineering-economic systems	2013
Brigadier General (Retired) Michael L. McGinnis	Systems and industrial engineering	2012
Dr. Andrew G. Loerch	Engineering and operations research	2011
Mr. Andrew W. Marshall	Economics	2010
Dr. Stuart H. Starr	Electrical engineering	2009
Dr. Thomas L. Allen	Economics and operations research	2008
Mr. Walter F. Hollis	Mathematics, physics	2007
Dr. Roy E. Rice	Operations research	2006
Dr. Vernon Bettencourt	Operations research	2005
General David M. Maddox	Operations research	2004
Ms. Natalie W. Crawford	Mathematics	2003
Mr. Michael F. Bauman	Engineering	2002
Dr. William G. Lese	Mathematical statistics, computer science	2001
Mr. Seymour J. Deitchman	Engineering	2000
Dr. Donald B. Rice	Engineering, economics	1999
General Larry D. Welch	Business administration, international relations	1998
Dr. Paul K. Davis	Chemical physics	1997
Dr. Edward C. Brady	Electrical engineering, mathematical economics	1996
Mr. E. B. Vandiver III	Physics	1995
Mr. James N. Bexfield	Mathematics, operations research	1994
Dr. David S. C. Chu	Economics	1993

Table C.2—Continued

Name	Disciplinary Background	Date
Major General John D. Robinson	Business administration	1992
Dr. Marion L. Williams	Engineering	1991
Mr. John K. Walker	Education	1990
Mr. Wayne P. Hughes	Operations research, mathematical statistics	1989
Clayton J. Thomas	Mathematics	1988
Wilbur B. Payne	Physics	1987
Dr. Seth Bonder	Industrial engineering	1986
Marion R. Bryson	Mathematics	1985
David A. Schrady	Operations research	1984
Major General Jasper A. Welch, Jr.	Physics	1983
Mr. Walter L. Deemer	Mathematics (imputed)	1982
Dr. Jack R. Borsting	Mathematics	1981
Lieutenant General Glenn A. Kent	Meteorology, mathematics, physics	1980
Dr. Bernard O. Koopman	Mathematics	1979
Dr. Philip M. Morse	Physics	1978

NOTES: The Vincent Wanner Award is the most prestigious award of the Military Operations Research Society. The disciplinary backgrounds were obtained via web searches and should not be considered definitive.

**Figure C.1
Disciplinary Background of RAND PhD Staff**

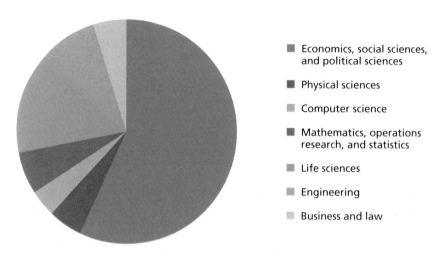

- Economics, social sciences, and political sciences
- Physical sciences
- Computer science
- Mathematics, operations research, and statistics
- Life sciences
- Engineering
- Business and law

SOURCE: Calculated using data from RAND Corporation, undated.
RAND *RR1469-C.1*

**Figure C.2
Disciplinary Background of the Institute for Defense Analyses Staff**

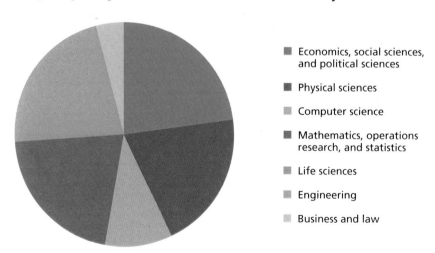

- Economics, social sciences, and political sciences
- Physical sciences
- Computer science
- Mathematics, operations research, and statistics
- Life sciences
- Engineering
- Business and law

SOURCE: Calculated using data from Institute for Defense Analyses, undated.
RAND *RR1469-C.2*

Options for Model Development, Maintenance, and Usage

This report calls for the reinvigoration of DoD's capabilities for joint modeling and analysis, with a mix of model types, human-gaming, and other research. A question that arose repeatedly during the study was where the locus of such work would be. Originally, the context was how to restore the campaign-modeling that existed in 2011, in CAPE's SAC, but the scope is now much broader. Some of the possibilities considered in the study were

1. none (there would be no DoD-coordinated function, except possibly for joint data collection and maintenance)
2. in the government (CAPE, J-8, a new entity in National Defense University, Net Assessment, or other)
3. FFRDCs.

Under any of these, commercial contractors might be used for some developments.

The first option is clearly feasible and, to a considerable extent, represents the status quo: Components award contracts as they see necessary, whether to private contractors or FFRDCs. Assuming that joint data remain an SSA function (or a function of the new ASSP), the approach would probably continue to suffice, although with inefficiencies.

Several options for in-government leads exist, as indicated, but it seems unlikely that CAPE or the Joint Staff will take on the responsibility without significant increases in billets and funding, as well as persuasive arguments about the need to do so. Creating a new entity in the National Defense University would be a risky endeavor with a long buildup time; Net Assessment would be a poor choice for the same reason and because the office has traditionally had very different talents and niches.[1] A primary consideration in this judgment is that developing the necessary analytic talent and organization is a substantial challenge. *High-quality* modelers and analysts cannot be quickly found and hired, and a high-quality organization to perform these functions simply could not arise overnight. Because the in-government options did not appear attractive, I gave more consideration to FFRDC options.

Ad Hoc or with Coordination by ASSP

For some years, DoD components have used FFRDCs on an ad hoc basis for a wide range of studies on broad and focused strategic research, particular military challenges, and particular examples of modeling or human wargaming. The default option is simply to continue. Although the term *ad hoc* can be pejorative, that is not intended here. This option would not require any new administrative apparatus. If an ASSP is created, then one of its functions could be to coordinate arrangements with the FFRDCs (with CAPE and the Joint Staff taking the lead). It would be essential to ensure that the FFRDCs undertaking the work understood the spirit intended, as discussed throughout this report in connection with the ASSP. Otherwise, the result might have an FFRDC label but would have the same shortcomings as the earlier campaign-modeling that CAPE chose to dissolve in 2011.

A WSEG Model

An alternative approach would be more highly structured. From 1948 to 1976, the Institute for Defense Analyses had a division called the Weapon Systems Evaluation Group (WSEG) attached to the Joint Staff and usually headed by a serving three-star military officer (Ponturo,

1979). WSEG's studies were almost all for the Joint Chiefs of Staff or the Director for Defense Research and Engineering, who was the highest official under the secretary and deputy secretary. The first director was Philip M. Morse, who returned to his MIT professorship in 1950. The final director was Lt Gen Glenn Kent (U.S. Air Force). WSEG was a wholly in-house organization with about 50 professionals, half military and half civilian. WSEG was able to hire top-notch talent without the then-stringent restrictions on salary. In the early years, WSEG played a preeminent role in DoD analysis. That changed in the 1960s, with the emergence of such offices as OSD's Systems Analysis (later PA&E and CAPE), other FFRDCs (e.g., RAND, Center for Naval Analyses, Research Analysis Corporation, Aerospace, MITRE), and the greatly increased analytic capabilities in the services and the Joint Chiefs of Staff. The Secretary of Defense disestablished WSEG in 1976, writing: "It is no longer needed . . . given the extensive complex of study and evaluation activities available to the Department" (Ponturo, 1979).

A Distributed WSEG Model

An alternative approach modernizing from the WSEG experience might be more suitable today. It would use two or more FFRDCs for different facets of support to joint analysis. With modern networking, the FFRDCs could be functionally well integrated. FFRDCs have different relative strengths and different relative interests in various types of analysis. Those can change over time. In practice, something like this (using multiple FFRDCs) is the situation today, but without any special organizational or legal relationships. Different DoD components ask their FFRDCs for particular studies or activities on an ad hoc basis. Congress also mandates some studies that are ultimately accomplished by FFRDCs (including this report).[2] The activities are not, however, coordinated, much less integrated.

A possible advantage to a "modern WSEG," at least for niche purposes, would be that it might make it far easier to have a stable group of top-flight analysts who were read into the various compartmented programs on which independent analysis is needed. That would require significant early effort, but the result might then be maintainable and efficient. Dealing with compartmented information has become a major problem over the past decade or so, within the DoD itself and FFRDCs.

Notes

Notes to Summary

[1] The words have distinct meanings here. They refer to *flexibility* across missions, *adaptiveness* to different circumstances, and *robustness* to shocks. Authors often use one of these words as shorthand for all three, as in exhorting planning for adaptiveness or seeking robust capabilities. The 2010 Quadrennial Defense Review also used the word *versatility* in this way. Still other authors use the term *agility*.

[2] The elements of the list relate to what DoD variously refers to as *challenges, illustrative planning scenarios, integrated security constructs,* and *defense-planning scenarios*.

[3] *Low resolution* can be illustrated with two examples: (1) a strategic-nuclear scenario specifies the alert state of U.S. forces and the type of retaliatory attack for which it is ready; (2) a strategic-mobility scenario requires airlift capacity as measured in millions of ton-miles per day, without specifying what is lifted or where it goes. Such low-resolution depictions should be approximately consistent with more-detailed descriptions.

Notes to Chapter One

[1] This research adhered to policies approved by RAND and DoD for the protection of human subjects. Those DoD officials interviewed provided their personal views and did not represent any position of DoD.

[2] The materials included overview briefings of the SSA activity, completed analytic baselines, current memoranda describing in-progress work, briefings sent to high officials, and published papers by interviewees. Those interviewed were chosen to ensure hearing from personnel in, or previously in, each of the major offices in the Office of the Secretary of Defense (OSD), the Joint Staff, and the services. By the midpoint of interviewing, very few new issues and points of disagreement were being mentioned. Instead, interviewees were presenting examples and nuances

reflecting their personal experiences. The review process served as another check to ensure that the most consequential issues and disagreements had been heard.

[3] Campaign analysis studies a phase of war that has a series of operations to accomplish military objectives. Two illustrative campaigns were the invasion of Iraq in 2003 and the surge in Iraq in 2007. Campaign analysis for hypothetical scenarios is used to evaluate future force structures and capability mixes. Scenarios may involve concurrent crises across the globe. Campaign analysis may be accomplished with complex computer simulation models, simpler models, and human wargaming.

[4] The services retain these capabilities and federally funded research and development centers (FFRDCs) often conduct such analysis for OSD, the Joint Staff, or services. Some officers on joint assignments can perform campaign analysis even if they are not using campaign models.

[5] The results of these models can be understood and explained by experts familiar with the models, but not necessarily easily. Such explanations are beyond the abilities of those who can run the model but do not appreciate underlying details or do work at code level. An expert analyst, if unfamiliar with model details or lacking the considerable time to deal with subtleties, may not find the model's behavior to be intuitive. Some interviewees said, bluntly, that they "did not trust" the models or data.

[6] The shift to capability-based planning was made in 2001 (Rumsfeld, 2001), presaged by the first QDR, which referred to being able to "respond to the full spectrum of crises" (Cohen, 1997). The shift had been urged in earlier books (e.g., Davis, 1994) and reports (e.g., National Defense Panel, 1997). As discussed elsewhere (Davis, 2014, p. 107), DoD's initial implementation had complex processes and led to serious misconceptions, such as the incorrect notion that the new approach did not use scenarios, name specific threats, or deal with resource constraints.

[7] The need for continued vigorous change has been recognized across administrations (Rumsfeld, 2006; Gates, 2010; Hagel, 2014)—most recently by Secretary Ashton Carter. The strategic issues have been described in terms of "eroding foundations of American power" (Krepinevich, 2009) and "looming discontinuities" (Davis and Wilson, 2011). To say the least, recent trends are not favorable (Ochmanek, 2014).

[8] Early materials included Program Analysis and Evaluation (1979), a review of base force development (Jaffe, 1993), a review of the two-decade period of planning documents, starting in 1980 (Larson, Orletsky, and Leuschner,, 2001), and a recent update (Larson et al., forthcoming).

[9] These included the *Bottom Up Review* and subsequent QDRs (Aspin, 1993; Cohen, 1997; Rumsfeld, 2001; Rumsfeld, 2006; Gates, 2010; Hagel, 2014), as well as an implementation plan (Joint Defense Capabilities Study Team, 2004), Defense Science Board reports (Defense Science Board, 1996a, 1996b, 1998, 2003a, 2003b,

2009), and a study for the Assistant Secretary for Defense for Research and Engineering (Mullen, 2013).

[10] Certain National Academies of Sciences, Engineering, and Medicine reports were helpful (National Research Council, 1997, 2005, 2006). Some FFRDC studies addressed fundamentals (Davis, 2014), and some reports were about rethinking analytic methods (Allen et al., 2007; Davis and Henninger, 2007) or based on working directly with high officials (Davis, Shaver, and Beck, 2008).

[11] Many informal materials are in the archives of the Military Operations Research Society. The most-relevant materials described meetings before or after QDRs; these demonstrate efforts to take on new challenges (Bexfield and Disbrow, 2004; Allen and Bexfield, 2006; Patenaude and Stanic, 2010; Davis, 1998; Leonard, Bexfield, and Sharman, 1998; Leonard and Bexfield, 2000; Thomason and Dechant, 2008; Solly, 2014; Military Operations Research Society, 2013; Leonard, Thomas, and Bexfield, 2013). See also a briefing on British work (Solly, 2015).

[12] See, e.g., a report from the Technical Cooperation Program (Taylor, 2011).

[13] The best-known account, originally published in 1971, describes early systems analysis (Enthoven and Smith, 2005), calling on a decade's efforts for the Air Force (Hitch and McKean, 1960). Broader discussions appeared in edited volumes (Quade, 1966; Quade and Boucher, 1968; Miser and Quade, 1988; Quade and Carter, 1989). William Kaufmann, an advisor to several defense secretaries, wrote about force-structuring analysis (Kaufmann, 1981, 1986, 1991).

Notes to Chapter Two

[1] That SSA activities serve very different functions is discussed in Fitzsimmons, 2012, and Dechant et al., 2008. One theme of the latter is the need for unclassified variants of scenarios that can be used readily in educational settings. A theme of the former is that scenario needs are different for strategy development and program development. Fitzsimmons also argues for putting more relative emphasis on analysis rather than scenario development.

[2] This draws on Davis, 2014, which reflects ideas and methods developed specifically for two Under Secretaries for Acquisition, Technology and Logistics (AT&L). The context was capability-area reviews (Davis, Shaver, and Beck, 2008), but the approach applies equally well to Secretary of Defense issues affecting the size and capabilities of future forces.

[3] A senior leader will seldom have time to delve into details but may do so selectively—to test the quality of staff work or to better understand a crucial issue. Organizing decision-aiding so that real-time zooming is *possible* sharpens work even if zooming is invoked rarely.

[4] In private industry, the function is sometimes served by management consultants, such as McKinsey and Company or the Boston Consulting Group, reporting to the chief executive officer.

[5] The words have distinct meanings here and refer to *flexibility* across missions, *adaptiveness* to different circumstances, and *robustness* to shocks. Authors often use one word as shorthand for all three, as in exhorting planning for adaptiveness or seeking robust capabilities. The 2010 QDR used the word *versatility* in this way. Still other authors use the term *agility*, as in some of NATO's work (Alberts, 2011).

[6] See Davis, 2014. The new responsibility has long been desired, as suggested in a 1992 talk by the director of PA&E exhorting analysts to discuss *why* conclusions emerge, *how* results would change with different assumptions, *why* to believe that a particular system or capability is needed, *why* new and varied scenarios are needed, and *why* we need to plan forces that will be used in a wide variety of future situations (Chu, 1992).

[7] Usage is inconsistent, but *measure* refers to an amount or degree of something. A *metric* is a standard of measurement, such as the ratio between an amount and an amount regarded as adequate.

[8] As an example, circa-2000 guidance for air forces to plan capabilities for halting an advancing mechanized army meant little without indicating how big an army could be thwarted, how quickly, and whether air supremacy should be assumed.

[9] As one example, the armored-vehicle kills per day by air forces in a continuing campaign against an invader depend on lower-level component factors, such as sorties per day and kills per sortie, while kills per sortie depend on still-lower–level factors, such as kills per engagement and the number of engagements per sortie (a function of command and control). Improving the top-most measure of effectiveness might actually be achieved by *reducing* sorties per day while increasing engagements per sortie and kills per engagement. Merely flying more sorties per day might be counterproductive if it came at the expense of per-engagement effectiveness.

[10] The Secretary of Defense issues contingency planning guidance, approved by the President, for the Chairman of the Joint Chiefs of Staff to use in developing the Joint Strategic Capabilities Plan. OUSDP has the lead role in preparing the contingency planning guidance and reviewing subsequent plans. The legislation is United States Code, Title 10, Section 113 (g)(2).

[11] The battle for resources is fundamental for officers working in the Pentagon, as recounted in a classic book (Smith and Gerstein, 2007).

Notes to Chapter Three

[1] Campaign analysis can be accomplished without using the particular campaign models of earlier SSA, or indeed without any computer models, but that has been rare within DoD. Many of those interviewed for this study decried the reduced capability in OSD and the Joint Staff for campaign *analysis*, not just campaign-modeling.

[2] Some of those critical of SSA cited anecdotes in which SSA representatives based assertions on questionable assumptions, but the vast majority of those interviewed agreed that SSA's analytic baseline work was well conceived and managed. It included innovative efforts to use human wargaming and, in recent years, to apply social-science experts.

[3] This judgment reflects a RAND study for the U.S. Army, which concluded that SSA (or, more precisely, the efforts of the earlier analytic agenda activity) had proven markedly valuable for preparation of the 2010 QDR, much more so than for the 2006 QDR (Larson et al., forthcoming).

[4] More precisely, the senior analysis officials associated with SSA supported research funded through the Modeling and Simulation Coordination Office (MSCO). Published examples include Davis and Cragin, 2009; Davis, 2011; and a variety of classified reports. For an overview see Coulter, 2009. For discussion in terms of unrestricted warfare, see Akst, 2009. MSCO also sponsored and participated in related special meetings of the Military Operations Research Society. Such efforts continue, as illustrated in 2016 by special meetings on human wargaming.

[5] This is based on my discussions in the mid-1990s with the OSD officials Ted Warner (then Assistant Secretary of Defense [ASD] for Strategy and Resources) and Fred Frostic (in Warner's office), as well as more-recent discussions with the office.

[6] PPBE-related analysts tend to worry about cost-effectiveness and to be skeptical about how quickly new capabilities will emerge and how effective they will be. Analysts associated with research and development tend to be more technology-push oriented and to be optimistic about the effectiveness of future capabilities and the speed with which they could be attained. Both sides can point to supportive examples. Precision fires and precision navigation transformed U.S. military operations in the course of a decade but proved far less effective when Serbia's tanks hid within forests and when manpower-intensive operations were necessary in counterinsurgency. DoD needs healthy competition between the two views.

[7] Some compartmented programs are aggressive and forward-looking. Some in SSA leadership have the necessary clearances. It is often possible to conduct related analysis using modifications of mainstream SSA data and models. Reportedly, however, there are limitations in the ability to do fully joint work, as noted by a long-time SSA leader and chief Navy analyst, Trip Barber (Barber, 2015).

[8] This later became the Modeling and Simulation Coordination Office.

[9] The Multi-Service Force Deployments assume adversary characteristics based on intelligence projections. In one account, a Multi-Service Force Deployment can be a 500-page description of how the postulated conflict unfolds. The description includes data for orders of battle, strategy and tactics at the operational level of war, axes of attack, defensive dispositions, tables of equipment, force allocations to missions, operations tempo and sortie rates, readiness factors, munitions, and sustainment (Cerniglia-Mosher, 2009).

[10] The background research includes in-house work and sponsored research at FFRDCs (e.g., Lostumbo et al., 2013, 2016). Some of this has also helped to develop the defense planning scenarios.

[11] The models in question, known by their acronyms, have included TACWAR, THUNDER, JICM, and STORM. All are good models but complex. JICM was originally designed to be more strategic level in nature and be correspondingly simpler, but embellishments and practice have made it less so.

[12] SSA recognized this problem and supported some exceptional efforts. In particular, the Marine Corps Combat Development Center conducted an extensive study, the *Joint Irregular Warfare Analytic Baseline (JIWAB)* (Wong, 2015), that broadly approached a challenging intervention scenario with interagency representatives and social-science experts with varied conceptual frameworks and tools. The intent was to think through a CONOPS that would properly deal with the political, economic, and human issues often omitted in military analysis. The study was never formally turned into an analytic baseline, but it illustrated what the analytic community might do if unshackled from traditional methods and models. Other SSA efforts that did not involve warfighting campaign models included use of the British Peace Support Operations Model (PSOM) (Body and Marston, 2011) and looking at the problem of securing nuclear weapons in a chaotic environment.

[13] This point was made in the 2010 QDR (Gates, 2010), the development of which had included a good deal of scenario variation in much the same spirit as described here. That QDR made some use of SSA products (data and baselines) but was notable for the low-resolution analysis that addressed other issues and different combinations of stress on U.S. military capabilities. Significantly, that analysis depended on *expertise* available in CAPE's Simulation and Analysis Center because of the prior in-depth campaign analysis, even though the analysis made little direct use of the campaign models.

[14] In this report, *capability-based planning* means "[p]lanning, under uncertainty, to provide capabilities suitable for a wide range of modern-day challenges and circumstances while working within an economic framework that necessitates choice" (Davis, 2002, p. xi). In this context, *capability* covers what others refer to as *capabilities and capacities*. The term *capability-based planning* has baggage because of DoD's notoriously complex processes for its implementation and basic misunderstandings,

including a false contrast with threat-based planning. Capability-based planning makes extensive use of scenarios and considers both generic and specific threats. The confused history is described elsewhere (Davis, 2014).

[15] This can be seen, for example, in congressional testimony by a recent defense official about U.S. challenges in the western Pacific, which discusses numerous problems and the importance of the "third offset strategy" announced by Secretary Chuck Hagel (Ochmanek, 2014).

[16] Standard academic references include the work of Henry Mintzberg (Mintzberg, 1994; Mintzberg, Lampel, and Ahlstrand, 2005).

[17] Excellent use of optimization can be found in numerous past studies, by both DoD and a few foreign militaries, such as the British Defence Operational Analysis Establishment in supporting a major UK review (Hoehl and Scales, 2011).

[18] Various serving or past officials have lamented the consensus problem while making other suggestions regarding how to improve strategic planning in DoD (Flournoy, 2015; Hicks, 2014).

Notes to Chapter Four

[1] One version was presented as early as 2004 by Kenneth Krieg, then director of PA&E and later USD(AT&L) (Bexfield and Disbrow, 2004).

[2] Because of resource limitations, this study did not address current and near-term planning, which use similar methods but focus on current war plans and their extrapolation.

[3] The elements of the list relate to what DoD variously refers to as *challenges, illustrative planning scenarios, integrated security constructs*, and *defense-planning scenarios*.

[4] Historical examples include the introduction of precision fires and networking, armed aerial combat vehicles, and the scale-up of such unmanned operations. Looking forward, some examples might be found in the systems being examined in DoD's Strategic Capabilities Office (Carter, 2016). These reportedly include use of microdrone packs for surveillance and other functions, new concepts for infantry, new roles for unmanned fighters (the Loyal Wingman initiative), and arsenal aircraft (Lamothe, 2016).

[5] Some argue that having the Joint Staff take on this role is implausible, but it should be noted that during the 1990s, the Joint Staff was an important proponent of innovation, as reflected in its *Joint Vision 2010* document and other initiatives of the era (Joint Staff, 1996).

[6] Strategic considerations include overall force size, the portfolio of capabilities, and how to go about achieving various classes of capability.

[7] For the basic character, see DoD, 1992. Although its source is not shown, it is related to or part of the defense planning guidance for fiscal years 1994–1999.

[8] Not much information is readily available in the public literature about DoD's planning scenarios. One overview is quite useful in this regard (Allen, 2011), indicating the nature of both single-challenge scenarios and integrated security constructs that postulate multiple challenges overlapping in time (e.g., defeat a regional rogue, deal with a terrorist attack on the homeland, *and* deter other adversaries). See also Gates, 2010, pp. 41–43.

[9] The 2010 QDR considered variations stressing U.S. capabilities in diverse ways (Gates, 2010, pp. 41–43). The related analysis depended on quick and simple analysis, not campaign models or detailed scenarios. This analysis benefited greatly, however, from the some analytic baselines on the shelf and from the analytic expertise of those in CAPE's SAC who had worked with the SSA campaign models and scenarios. When the campaign-modeling was terminated, baselines were no longer developed, and expertise disappeared as well.

[10] See discussion and references cited in Davis, 2014.

[11] Low resolution can be illustrated with two examples: (1) a strategic-nuclear scenario specifies the alert state of U.S. forces and the type of retaliatory attack for which it is ready; (2) a strategic-mobility scenario requires airlift capacity as measured in millions of ton-miles per day, without specifying what is lifted or where it goes. Such low-resolution depictions should be approximately consistent with more-detailed descriptions.

[12] This assertion reflects discussions with the late Ted Warner, ASD for Strategy and Resources in the 1990s; other veterans of the office; and personal observations.

[13] Others have argued for much broader reasons that OSD would benefit from a dedicated decision-support cell focused on strategic issues (Lamb and Lachow, 2006).

[14] An early version appeared in a study sponsored by PA&E through the Defense Modeling and Simulation Office, which generated two reports (Davis and Henninger, 2007; Allen et al., 2007). The study pulled together ideas raised and debated in analytic-community workshops to suggest a master plan for analysis. A version of the first report was adopted as part of National Research Council, 2006. That version appeared in Davis, 2014.

[15] Ironically even with the recent emphasis on operational-level wargames, DoD still tends to focus on state-on-state wars, such as in Eastern Europe and the Pacific (e.g., Taiwan Straits, South China Sea). There has been relatively little study of ISIL and other types or irregular or unconventional wars.

[16] A large published volume summarized the initial work (Davis and Cragin, 2009). Work on public support for terrorism was then released (Davis, Larson, et al., 2012). It included new case studies testing validity of the qualitative models. The models passed this test in the context of providing explanation, structure, and useful but uncertain diagnosis and prescription. The models were not intended to be predictive.

[17] The earliest version was sponsored by the Office of Naval Research as part of DoD's Human Social Culture Behavior Modeling Program and by the Marine Corps (Davis and O'Mahony, 2013). This has led recently to research on *heterogeneous fusion*—i.e., fusion of different kinds of information from diverse sources and with such characteristics as being soft, ambiguous, contradictory, and even deceptive (Davis, Perry, et al., 2016).

[18] Variants of the definition appear in numerous sources, such as Lempert, Popper, and Bankes, 2003, a report on very-long-term planning.

[19] For a review of concepts dating back to the 1980s and national-security applications since the early 1990s, see Davis, 2012. For recent applications see also Lempert, Warren, et al., 2016.

[20] DoD has used the term *wargame* to mean "human wargame," whereas others have often used the term to include the use of computer models.

[21] The memorandum was issued in May 2015 (Work, 2015).

[22] See Beall, 2015, for discussion by a senior executive on the Navy staff with long experience in both naval and joint studies and wargaming.

[23] A well-known book on human wargaming was developed at the Center for Naval Analyses by Peter Perla (Perla, 1990) and has been recently updated (Curry, 2012). Many analysts who see value in wargaming, including this author, have been influenced over the years by commercial contributions, particularly by James Dunnigan (Dunnigan, 2003). The Office of Net Assessment has used human wargaming since its inception.

[24] One published example dealt with Palestine-Israel issues (Molander et al., 2009). Roger Molander was one of the primary figures in the Day After games, in addition to Peter Wilson of RAND.

[25] For another discussion of how to relate gaming back to analysis, see Perla and McGrady, 2009.

[26] The model-test-model approach is common in some kinds of high-technology industry, including development of advanced missiles. Since it is not possible to test such systems exhaustively in the field, decisions must depend on believing the model of the system. Field tests, then, are designed to test or provide information on aspects on the model that are most in need of empirical information.

[27] This work is being led under CAPE sponsorship by Joel Predd and Igor Mikolic Torreira of RAND. The intent is "exploratory force-structure analysis." No unclassified documentation exists as yet.

[28] Most analysts in CAPE frame issues, review other analyses, and sometimes do back-of-the-envelope calculations to inform decisions. Much of these analysts' time is spent on programmatic issues—defining the issues, leading program review teams, dealing with cost estimates and budgets, etc. Service analysts are much more likely to build and use models to perform detailed studies (e.g., analysis of alternatives).

Note to Appendix A

[1] The author confirmed in discussions with staff of the House Armed Services Committee that the primary intent of the study was to assess the SSA activity.

Notes to Appendix B

[1] Intent was reflected by the Principal Under Secretary of Defense (Policy), Ryan Henry, in an address to the Military Operations Research Society in which he said that the directive would "link DoD-wide decisions with the defense strategy and apportion risk across the range of challenge areas: traditional, irregular, catastrophic, and disruptive" (quoted in Bexfield and Disbrow, 2004, p. 8).

[2] Little of this occurred, although two such analytic baselines were developed by U.S. Pacific Command.

Note to Appendix C

[1] The data at the Institute for Defense Analyses website (Institute for Defense Analyses, undated) have been mapped approximately into the same structure as that for RAND.

Notes to Appendix D

[1] The Net Assessment Office has remained outside the fray of the PPBE and acquisition processes, although it contributes research and analysis. It has also tended not to be especially *analytic* in the sense of that term used in SSA circles: The office has often emphasized softer methods, such as studying technological trends, human wargaming, applying the concepts of strategic competition, and thinking creatively

about possible future scenarios (Krepinevich, 2010). Analytical models have played only a small role in its work. There were exceptions in the 1980s, but the office concluded subsequently that the other methods were more productive and efficient for its purposes.

[2] An example is the analysis of alternatives for aerial refueling (Kennedy, 2006).

Abbreviations

ASD	Assistant Secretary of Defense
ASSP	Analytic Support for Strategic Planning (proposed)
CAPE	Cost Assessment and Program Evaluation
CONOPS	concepts of operations
DoD	Department of Defense
FARness	flexibility, adaptiveness, and robustness
FFRDC	federally funded research and development center
JICM	Joint Integrated Contingency Model
M&S	modeling and simulation
MS&A	modeling, simulation, and analysis
MSFD	Multi-Service Force Deployment
NATO	North Atlantic Treaty Organization
OSD	Office of the Secretary of Defense
OUSDP	Office of the Under Secretary of Defense for Policy
PA&E	Office of Program Analysis and Evaluation

PPBE	planning, programming, budgeting, and executing
QDR	Quadrennial Defense Review
SAC	Simulation and Analysis Center
SSA	support for strategic analysis
USD(AT&L)	Office of the Under Secretary of Defense for Acquisition, Technology and Logistics
WSEG	Weapon Systems Evaluation Group

Bibliography

Akst, George, "Analysis Support for Unrestricted Warfare," *Proceedings on Combating the Unrestricted Warfare Threat: Terrorism, Resources, Economics, and Cyberspace*, Laurel, Md.: Johns Hopkins University Applied Physics Laboratory, 2009, pp. 417–423.

Alberts, Davis S., *The Agility Advantage: A Survival Guide for Complex Enterprises and Endeavors*, Washington, D.C.: DoD Command and Control Research Program, 2011.

Allen, Thomas, "Some Challenges for Global Military Operations," paper presented at the 8th Annual Disruptive Technologies Conference, Washington, D.C., November 8–9, 2011.

Allen, Thomas, Sheila B. Bankes, Jerome Bracken, Paul K. Davis, Jason A. Dechant, Olaf Elton, Michael F. Fitzsimmons, Amy E. Henninger, Royce Kneece, Stuart H. Starr, and Marin R. Stytz, *Foundation for an Analysis Domain Modeling and Simulation Business Plan*, Alexandria, Va.: Institute for Defense Analyses, December 2007.

Allen, Tom, and James Bexfield, *MORS Workshop: Capabilities-Based Planning II: Identifying, Classifying and Measuring Risk in a Post 9-11 World*, Military Operations Research Society, Arlington, Va., April 4–6, 2006.

Aspin, Les, *Report of the Bottom Up Review*, Washington, D.C.: Department of Defense, 1993.

Barber, Arthur H., "Joint Warfare Analysis: The Key to Shaping DoD's Future," *Phalanx*, Vol. 47, No. 1, March 2014a.

———, "Rethinking the Future Fleet," *Proceedings Magazine*, Vol. 140, No. 5, May 2014b.

Barber, Trip, "Meeting New Defense Analytical Challenges: An Industry Perspective," *Phalanx*, December 2015, pp. 75–78.

Beall, Virginia "Robbin," "Defense Innovation Through Wargaming," *Phalanx*, Vol. 48, No. 3, September 2015.

Bexfield, James, and Lisa Disbrow, "Capabilities Based Planning: The Road Ahead," paper presented at the MORS Workshop, *Capabilities-Based Planning: The Road Ahead*, Arlington, Va., October 19–21, 2004.

Body, Howard, and Colin Marston, "The Peace Support Operations Model: Origin, Development, Philosophy and Use," *Journal of Defense Modeling and Simulation*, Vol. 8, No. 4, 2011, pp. 69–77.

CAPE—*See* Office of Cost Assessment and Program Evaluation.

Carter, Ashton B., "Remarks Previewing the FY 2017 Defense Budget," Washington, D.C.: Department of Defense, 2016.

Cerniglia-Mosher, Mary, "Analysis Panel Analytic Framework and Analytic Agenda Link," briefing, U.S. Air Force, March 18, 2009. As of July 22, 2016: http://www.afams.af.mil/shared/media/document/AFD-090416-077.pdf

Chu, David, "What Is the Analyst's Responsibility to the Decision Maker," *Phalanx*, Vol. 25, No. 3, 1992.

Cohen, William, *Report of the Quadrennial Defense Review*, Washington, D.C.: Department of Defense, 1997.

Coulter, Eric, "Sponsor's Corner," *Phalanx*, Vol. 37, No. 1, March 2004, pp. 38–39.

———, "Analysis Support for the Interagency," in Ronald R. Luman, ed., *Unrestricted Warfare Symposium 2009*, Laurel, Md.: Johns Hopkins University Applied Physics Laboratory, 2009, pp. 167–192.

Curry, John, ed., *Peter Perla's the Art of Wargaming: A Guide for Professionals and Hobbyists*, Bristol, UK: History of Wargaming Project, 2012.

Davis, Paul K., *New Challenges in Defense Planning: Rethinking How Much Is Enough*, Santa Monica, Calif.: RAND Corporation, MR-400-RC, 1994.

———, "Report of Working Group: Overall Force Planning Concepts, in Lessons Learned from the QDR," briefing, Arlington, Va.: Military Operations Research Society, 1998.

———, *Analytic Architecture for Capabilities-Based Planning, Mission-System Analysis, and Transformation*, Santa Monica, Calif.: RAND Corporation, MR-1513-OSD, 2002. As of August 1, 2013: http://www.rand.org/pubs/monograph_reports/MR1513.html

———, ed., *Dilemmas of Intervention: Social Science for Stabilization and Reconstruction*, Santa Monica, Calif.: RAND Corporation, MG-1119-OSD, 2011. As of April 8, 2013: http://www.rand.org/pubs/monographs/MG1119.html

————, *Some Lessons from RAND's Work on Planning Under Uncertainty for National Security*, Santa Monica, Calif.: RAND Corporation, TR-1249-OSD, 2012. As of July 22, 2016:
http://www.rand.org/pubs/technical_reports/TR1249.html

————, *Analysis to Inform Defense Planning Despite Austerity*, Santa Monica, Calif.: RAND Corporation, RR-482-OSD, 2014. As of June 14, 2016:
http://www.rand.org/pubs/research_reports/RR482.html

Davis, Paul K., and Kim Cragin, eds., *Social Science for Counterterrorism: Putting the Pieces Together*, Santa Monica, Calif.: RAND Corporation, MG-849-OSD, 2009. As of June 14, 2016:
http://www.rand.org/pubs/monographs/MG849.html

Davis, Paul K., David C. Gompert, Stuart Johnson, and Duncan Long, *Developing Resource-Informed Strategic Assessments and Recommendations*, Santa Monica, Calif.: RAND Corporation, MG-703-JS, 2008. As of April 8, 2013:
http://www.rand.org/pubs/monographs/MG703.html

Davis, Paul K., and Amy Henninger, *Analysis, Analysis Practices, and Implications for Modeling and Simulation*, Santa Monica, Calif.: RAND Corporation, OP-176-OSD, 2007. As of August 8, 2013:
http://www.rand.org/pubs/occasional_papers/OP176.html

Davis, Paul K., Eric Larson, Zachary Haldeman, Mustafa Oguz, and Yashodhara Rana, *Understanding and Influencing Public Support for Insurgency and Terrorism*, Santa Monica, Calif.: RAND Corporation, MG-1122-OSD, 2012. As of April 8, 2013:
http://www.rand.org/pubs/monographs/MG1122.html

Davis, Paul K., Jimmie McEver, and Barry Wilson, *Measuring Interdiction Capabilities in the Presence of Anti-Access Strategies: Exploratory Analysis to Inform Adaptive Strategies for the Persian Gulf*, Santa Monica, Calif.: RAND Corporation, MR-1471-AF, 2002. As of February 4, 2016:
http://www.rand.org/pubs/monograph_reports/MR1471.html

Davis, Paul K., and Angela O'Mahony, *A Computational Model of Public Support for Insurgency and Terrorism: A Prototype for More General Social-Science Modeling*, Santa Monica, Calif.: RAND Corporation, TR-1220-OSD, 2013. As of August 8, 2013:
http://www.rand.org/pubs/technical_reports/TR1220.html

Davis, Paul K., Walter L. Perry, John S. Hollywood, and David Manheim, *Uncertainty-Sensitive Heterogeneous Information Fusion: Assessing Threat with Soft, Uncertain, and Conflicting Evidence*, Santa Monica, Calif.: RAND Corporation, RR-1200-NAVY, 2016. As of June 14, 2016:
http://www.rand.org/pubs/research_reports/RR1200.html

Davis, Paul K., Russell D. Shaver, and Justin Beck, *Portfolio-Analysis Methods for Assessing Capability Options*, Santa Monica, Calif.: RAND Corporation, MG-703-JS, 2008. As of August 11, 2013:
http://www.rand.org/pubs/monographs/MG703.html

Davis, Paul K., and Peter A. Wilson, *Looming Discontinuities in U.S. Military Strategy and Defense Planning*, Santa Monica, Calif.: RAND Corporation, OP-326-OSD, 2011. As of August 8, 2013:
http://www.rand.org/pubs/occasional_papers/OP326.html

Dechant, Jason, James S. Thomason, Michael F. Fitzsimmons, Michael F. Niles, and Zachary S. Rabold, *Open Scenario Study, Phase 1*, Vol. 1: *Assessment Overview and Results*, Alexandria, Va.: Institute for Defense Analyses, 2008.

Defense Science Board, "Future Analysis Requirements and Related Needs for M&S," in *Tactics and Technology for 21st Century Military Superiority*, Vol. 2, Part 1: *Supporting Materials*, Washington, D.C.: Office of the Under Secretary of Defense for Acquisition, Technology and Logistics, 1996a, pp. 47–72.

———, *Tactics and Technologies for 21st Century Military Superiority*, Washington, D.C.: Office of the Under Secretary of Defense for Acquisition, Technology and Logistics, 1996b.

———, *Joint Operations Superiority in the 21st Century: Integrating Capabilities Underwriting Joint Vision 2010 and Beyond*, Washington, D.C.: Office of the Under Secretary of Defense for Acquisition, Technology and Logistics, 1998.

———, *Phase I Report of the Defense Science Board Task Force on Joint Experimentation*, Washington, D.C.: Office of the Under Secretary of Defense for Acquisition, Technology and Logistics, 2003a.

———, *The Role and Status of DoD Red Teaming Activities*, Washington, D.C.: Department of Defense, 2003b.

———, *Report of the 2008 Defense Science Board Study on Capability Surprise*, Vol. 1: *Main Report*, Washington, D.C., 2009.

Department of Defense, *Annex A: Illustrative Planning Scenarios*, Washington, D.C.: May 2, 1992, declassified. As of July 22, 2016:
http://www.archives.gov/declassification/iscap/pdf/2008-003-doc12.pdf

Department of Defense Directive 8260.05, *Support for Strategic Analysis (SSA)*, Washington, D.C.: Department of Defense, July 7, 2011.

Department of Defense Directive 8260.1, *Data Collection, Development, and Management in Support of Strategic Analysis*, Washington, D.C.: Department of Defense, December 6, 2002.

Department of Defense Directive 8260.1, *Implementation of Data Collection, Development, and Management for Strategic Analyses*, Washington, D.C.: Department of Defense, January 21, 2003.

DoD—*See* Department of Defense.

Drucker, Peter F., *The Effective Executive: The Definitive Guide to Getting the Right Things Done*, New York: HarperBusiness, January 3, 2006.

Dunnigan, James F., *How to Make War: A Comprehensive Guide to Modern Warfare in the Twenty-First Century*, 4th ed., New York: Quill, 2003.

Enthoven, Alain, and K. Wayne Smith, *How Much Is Enough: Shaping the Defense Program, 1961–1969*, Santa Monica, Calif.: RAND Corporation, CB-403, 2005.

Fitzsimmons, Michael, "A Proposal to Overhaul Support to Strategic Analysis," Alexandria, Va.: Institute for Defense Analyses, unpublished manuscript, 2012.

Flournoy, Michele A., *The Urgent Need for Defense Reform*, testimony before the Senate Armed Services Committee, Washington, D.C., December 8, 2015.

Gates, Robert, *Report of the Quadrennial Defense Review*, Washington, D.C.: Department of Defense, 2010.

Hagel, Chuck, *Quadrennial Defense Review 2014*, Washington, D.C.: Department of Defense, 2014.

Hicks, Kathleen, "Commentary: 21st Century Strategic Planning: U.S. Needs Agile, Focused Process for Modern Challenges," *Defense News*, April 21, 2014.

Hitch, Charles J., and Roland N. McKean, *Economics of Defense in the Nuclear Age*, Cambridge, Mass.: Harvard University Press, 1960.

Hoehl, Louise, and Thomas Scales, "Informing the Strategic Defence and Security Review: The Strategic Balance of Investment Study," Defence Science and Technology Laboratory, 2011.

Institute for Defense Analyses, "Our Research Staff," web page, undated. As of June 23, 2016:
https://www.ida.org/en/AboutIDA/OurResearchStaff.aspx

Jaffe, Lorna S., *The Development of the Base Force 1989–1992*, Washington, D.C.: Joint History Office, Office of the Chairman of the Joint Chiefs of Staff, 1993.

Joint Defense Capabilities Study Team, *Joint Defense Capabilities Study: Improving DoD Strategic Planning, Resourcing and Execution to Satisfy Joint Capabilities*, Washington, D.C.: Department of Defense, 2004.

Joint Staff, *Joint Vision 2010*, Washington, D.C.: Department of Defense, 1996.

Kaufmann, William W., *Defense in the 1980s*, Washington, D.C.: Brookings Institution Press, 1981.

———, *A Reasonable Defense*, Washington, D.C.: Brookings Institution Press, 1986.

———, *Decisions for Defense: Prospects for a New Order*, Washington, D.C.: Brookings Institution Press, 1991.

Keener, Ross, *Support for Strategic Analysis SSA 101—An Overview*, briefing, Washington, D.C.: U.S. Air Force and CAPE, September 2011

Kennedy, Michael, *Analysis of Alternatives (AoA) for KC-135 Recapitalization: Executive Summary*, Santa Monica, Calif.: RAND Corporation, MG-495-AF, 2006. As of June 15, 2016:
http://www.rand.org/pubs/monographs/MG495.html

Krepinevich, Andrew F., "The Pentagon's Wasting Assets: The Eroding Foundations of American Power," *Foreign Affairs*, July/August 2009, pp. 18–33.

———, *7 Deadly Scenarios: A Military Futurist Explores the Changing Face of War in the 21st Century*, New York: Bantam Books, 2010.

Krieg, Kenneth, "Capabilities Based Planning: The View from PA&E," briefing slides presented at the MORS Workshop, *Capabilities-Based Planning: The Road Ahead*, Arlington, Va., October 19–21, 2004.

Lamb, Christopher J., and Irving Lachow, *Reforming Pentagon Decisionmaking*, Washington, D.C.: Institute for National Strategic Studies, National Defense University, Strategic Forum, No. 221, 2006.

Lamothe, Dan, "Veil of Secrecy Lifted on Pentagon Office Planning 'Avatar' Fighters and Drone Swarms," *Washington Post*, March 8, 2016.

Larson, Eric V., Derek Eaton, Michael E. Linick, John E. Peters, Agnes Gereben Schaefer, Keith Walters, Stephanie Young, H. G. Massey, and Michelle Darrah Ziegler, *Defense Planning in a Time of Conflict: A Comparative Analysis of the 2001–2014 Quadrennial Defense Reviews, and Implications for the Army*, Santa Monica, Calif.: RAND Corporation, RR-1309-A, forthcoming.

Larson, Eric V., David T. Orletsky, and Kristin J. Leuschner, *Defense Planning in a Decade of Change: Lessons from the Base Force, Bottom-Up Review, and Quadrennial Defense Review*, Santa Monica, Calif.: RAND Corporation, MR-1387-AF, 2001. As of July 22, 2016:
http://www.rand.org/pubs/monograph_reports/MR1387.html

Lempert, Robert J., David G. Groves, Steven W. Popper, and Steven C. Bankes, "A General Analytic Method for Generating Robust Strategies and Narrative Scenarios," *Management Science*, Vol. 4, April 2006, pp. 514–528.

Lempert, Robert J., Steven W. Popper, and Steven C. Bankes, *Shaping the Next One Hundred Years: New Methods for Quantitative Long-Term Policy Analysis*, Santa Monica, Calif.: RAND Corporation, MR-1626-RPC, 2003. As of July 28, 2016:
http://www.rand.org/pubs/monograph_reports/MR1626.html

Lempert, Robert J., Drake Warren, Ryan Henry, Robert W. Button, Jonathan Klenk, and Kate Giglio, *Defense Resource Planning Under Uncertainty: An Application of Robust Decision Making to Munitions Mix Planning*, Santa Monica, Calif.: RAND Corporation, RR-1112-OSD, 2016. As of June 15, 2016:
http://www.rand.org/pubs/research_reports/RR1112.html

Leonard, Michael, and James Bexfield, *2000 Joint Analysis: QDR 2001 and Beyond*, briefing, Arlington, Va.: Military Operations Research Society, 2000.

Leonard, Michael, James Bexfield, and Peter Sharman, *QDR Analysis: Lessons Learned and Future Directions Mini Symposium*, briefing, Alexandria, Va.: Military Operations Research Society, 1998.

Leonard, Michael, Jim Thomas, and James Bexfield, *QDR: Role of Analytics in Addressing the New Budget Environment (Out Brief)*, briefing, Arlington, Va.: Military Operations Research Society, 2013.

Lostumbo, Michael, "A New Taiwan Strategy to Adapt to PLA Precision Strike Capabilities," in Roger Cliff, Phillip C. Saunders, and Scott Harold, *New Opportunities and Challenges for Taiwan's Security*, Santa Monica, Calif.: RAND Corporation, CF-279-OSD, 2011, pp. 127–149. As of July 29, 2016:
http://www.rand.org/pubs/conf_proceedings/CF279.html

Lostumbo, Michael J., David R. Frelinger, James Williams, and Barry Wilson, *Air Defense Options for Taiwan: An Assessment of Relative Costs and Operational Benefits*, Santa Monica, Calif.: RAND Corporation, RR-1051-OSD, 2016. As of July 28, 2016:
http://www.rand.org/pubs/research_reports/RR1051.html

Lostumbo, Michael J., Michael J. McNerney, Eric Peltz, Derek Eaton, David R. Frelinger, Victoria A. Greenfield, John Halliday, Patrick Mills, Bruce R. Nardulli, Stacie L. Pettyjohn, Jerry M. Sollinger, and Stephen M. Worman, *U.S. Overseas Military Presence: Relative Costs and Strategic Benefits*, Santa Monica, Calif.: RAND Corporation, RR-201-OSD, 2013. As of July 28, 2016:
http://www.rand.org/pubs/research_reports/RR201.html

Military Operations Research Society, *MORS QDR 2012 Workshop Synthesis Group*, briefing, Arlington, Va.: Military Operations Research Society, 2013.

Mintzberg, Henry, *Rise and Fall of Strategic Planning*, New York: Free Press, 1994.

Mintzberg, Henry, Joseph Lampel, and Bruce Ahlstrand, *Strategy Safari: A Guided Tour Through the Wilds of Strategic Management*, New York: Free Press, 2005.

Miser, Hugh J., and Edward S. Quade, eds., *Handbook of Systems Analysis*, New York: North Holland Publishing Company, 1988.

Molander, Roger C., David Aaron, Robert E. Hunter, Martin C. Libicki, Douglas Shontz, and Peter A. Wilson, *The Day After . . . in Jerusalem: A Strategic Planning Exercise on the Path to Achieving Peace in the Middle East*, Santa Monica, Calif.: RAND Corporation, CF-271-CMEPP, 2009. As of June 15, 2016:
http://www.rand.org/pubs/conf_proceedings/CF271.html

Mullen, Frank E., *Dynamic Multilevel Modeling Framework Phase I—Feasibility*, Alexandria, Va.: Modeling and Simulation Coordination Office, Department of Defense, 2013.

National Defense Panel, *Transforming Defense: National Security in the 21st Century*, Washington, D.C.: Department of Defense, 1997.

National Research Council, *Modeling and Simulation*, Vol. 9: *Technology for the United States Navy and Marine Corps: 2000–2035*, Washington, D.C.: Naval Studies Board, National Academies Press, 1997.

———, *Naval Analytical Capabilities: Improving Capabilities-Based Planning*, Washington, D.C.: Naval Studies Board, National Academies Press, 2005.

———, *Defense Modeling, Simulation, and Analysis: Meeting the Challenge*, Washington, D.C.: National Academies Press, 2006.

———, *U.S. Air Force Strategic Deterrence Analytic Capabilities: An Assessment of Methods, Tools, and Approaches for the 21st Century Security Environment*, Washington, D.C.: National Academies Press, 2014.

Ochmanek, David A., *The Role of Maritime and Air Power in DoD's Third Offset Strategy*, Santa Monica, Calif.: RAND Corporation, CT-420, 2014. As of June 15, 2016:
http://www.rand.org/pubs/testimonies/CT420.html

Ochmanek, David A., Edward Harshbarger, David E. Thaler, and Glenn A. Kent, *To Find, and Not to Yield: How Advances in Information and Firepower Can Transform Theater Warfare*, Santa Monica, Calif.: RAND Corporation, MR-958-AF, 1998. As of August 8, 2013:
http://www.rand.org/pubs/monograph_reports/MR958.html

Office of Cost Assessment and Program Evaluation, "Introduction," web page, undated. As of June 23, 2016:
http://www.cape.osd.mil/content/AboutCape.html

PA&E—*See* Program Analysis and Evaluation.

Patenaude, Annie, and Cy Stanic, *MORS Analytic Agenda Conference*, briefing, Arlington, Va.: Military Operations Research Society, 2010.

Perla, Peter P., *The Art of Wargaming: A Guide for Professionals and Hobbyists*, Annapolis, Md.: Naval Institute Press, 1990.

Perla, Peter P., and Ed McGrady, *Systems Thinking and Wargaming*, Alexandria, Va.: Center for Naval Analyses, 2009.

Perry, William J., and John P. Abizaid, *Ensuring a Strong U.S. Defense for the Future: The National Defense Panel Review of the 2014 Quadrennial Defense Review*, Washington, D.C.: U.S. Institute for Peace, 2014.

Ponturo, John, *Analytical Support for the Joint Chiefs of Staff: The WSEG Experience, 1948–1976*, Alexandria, Va.: Institute for Defense Analyses, 1979.

Program Analysis and Evaluation, *Capabilities for Limited Contingencies in the Persian Gulf*, Washington, D.C.: Office of the Secretary of Defense, 1979.

Quade, Edward S., ed., *Analysis for Military Decisions*, Chicago: RAND McNally; Amsterdam: North-Holland, 1966.

Quade, Edward S., and Wayne I. Boucher, eds., *Systems Analysis and Policy Planning: Applications for Defense*, New York: Elseveier Science Publishers, 1968.

Quade, Edward S., and Grace M. Carter, eds., *Analysis for Public Decisions*, 3rd ed., New York: North Holland Publishing Company, 1989.

RAND Corporation, "RAND at a Glance," web page, undated. As of July 11, 2016:
http://www.rand.org/about/glance.html

Rumsfeld, Donald, *Report of the Quadrennial Defense Review*, Washington, D.C.: Department of Defense, 2001.

———, *Quadrennial Defense Review Report*, Washington, D.C.: Department of Defense, 2006.

Shlapak, David A., and Michael W. Johnson, *Reinforcing Deterrence on NATO's Eastern Flank: Wargaming the Defense of the Baltics*, Santa Monica, Calif.: RAND Corporation, RR-1253-A, 2016. As of June 15, 2016:
http://www.rand.org/pubs/research_reports/RR1253.html

Smith, Perry McCoy, and Daniel M. Gerstein, *Assignment: Pentagon: How to Excel in a Bureaucracy*, Dulles, Va.: Potomac Books, 2007.

Solly, Rob, *The Use of Analysis in UK Defense Reviews*, briefing, Arlington, Va.: Military Operations Research Society, 2014.

———, "Expressing Deep Uncertainty as Levels of Understanding," briefing, Decisionmaking Under Uncertainty Conference, Delft, Netherlands, 2015.

Stevens, Jim, *Perspectives on the Analysis M&S Community*, briefing, Washington, D.C.: Office of the Secretary of Defense, Program Analysis and Evaluation, 2008.

Taylor, Ben, *TTCP Technical Report: Analysis Support to Strategic Planning*, Ottawa: The Technical Cooperation Program and Defense Research and Development, 2011.

Thomason, Jim, and Jason Dechant, *Strengthening the Next QDR Through Timely and Relevant Analysis: Working Group 5; Strategy, Force, and Program Integration*, briefing, Arlington, Va.: Military Operations Research Society, 2008.

Tirpak, John A., "The New Air Force Program," *Air Force Magazine Online*, Vol. 89, No. 7, 2006.

United States Code, Title 10, Section 113, Secretary of Defense, Washington, D.C., January 3, 2012.

U.S. Congress, Carl Levin and Howard P. "Buck" McKeon National Defense Authorization Act for Fiscal Year 2015, Washington, D.C., Public Law 113-291, December 19, 2014.

Wong, Yuna Huh, "Preparing for Contemporary Analytic Challenges," *Phalanx*, Vol. 47, No. 4, December 2014, pp. 35–39.

———, *Joint Irregular Warfare Analytic Baseline (JIWAB)*, briefing, Quantico, Va.: Marine Corps Combat Development Command, 2015.

Wong, Yuna, and Sara Cobb, *Narrative Analysis in Seminar Gaming*, briefing, Quantico, Va.: Marine Corps Combat Development Command, 2014.

Work, Robert O., Deputy Secretary of Defense, "Wargaming Summit Way Ahead," memorandum, May 8, 2015.